Resource Rich Muslim Countries and Islamic Institutional Reforms

ECONOMIC, POLITICAL AND SOCIAL INSTITUTIONS IN ISLAM

Hossein Askari and Dariush Zahedi
Series Editors

Vol. 1

The Economic, Political and Social Institutions in Islam series
is part of the Peter Lang Politics and Economics list.
Every volume is peer reviewed and meets
the highest quality standards for content and production.

PETER LANG
New York • Bern • Berlin
Brussels • Vienna • Oxford • Warsaw

Liza Mydin, Hossein Askari
and Abbas Mirakhor

Resource Rich Muslim Countries and Islamic Institutional Reforms

PETER LANG
New York • Bern • Berlin
Brussels • Vienna • Oxford • Warsaw

Library of Congress Cataloging-in-Publication Data

Names: Mydin, Liza, author. | Askari, Hossein, author. | Mirakhor, Abbas, author.
Title: Resource rich Muslim countries and Islamic institutional reforms /
Liza Mydin, Hossein Askari and Abbas Mirakhor.
Description: New York: Peter Lang, 2018.
Series: Economic, political and social institutions in Islam; vol. 1
ISSN 2573-2161 (print) | ISSN 2573-217X (online)
Includes bibliographical references and index.
Identifiers: LCCN 2017027971 | ISBN 978-1-4331-4351-9 (hardback: alk. paper)
ISBN 978-1-4331-4352-6 (ebook pdf) | ISBN 978-1-4331-4353-3 (epub)
ISBN 978-1-4331-4354-0 (mobi)
Subjects: LCSH: Organisation of Islamic Cooperation.
Natural resources—Islamic countries. | Resource curse—Islamic countries.
Petroleum industry and trade—Islamic countries. | Islamic countries—Economic conditions.
Economics—Religious aspects—Islam.
Classification: LCC HC499.Z65 M93 2017 | DDC 333.70917/67—dc23
LC record available at https://lccn.loc.gov/2017027971
DOI 10.3726/b11157

Bibliographic information published by **Die Deutsche Nationalbibliothek**.
Die Deutsche Nationalbibliothek lists this publication in the "Deutsche
Nationalbibliografie"; detailed bibliographic data are available
on the Internet at http://dnb.d-nb.de/.

The paper in this book meets the guidelines for permanence and durability
of the Committee on Production Guidelines for Book Longevity
of the Council of Library Resources.

29 Broadway, 18th floor, New York, NY 10006
www.peterlang.com

Printed in Germany

To our families

CONTENTS

ILLUSTRATIONS

Figure

Tables

FOREWORD

Latifah Merican Cheong

There have been numerous studies on the resource curse. This book is a first on this subject from a new perspective and focus of the resource curse in Muslim countries endowed with natural resources. The authors have done extensive economic analysis to trace the developments in the commodities-rich Muslim countries and provide the evidence-based explanations for their rather inferior economic performance. In many Muslim countries, the unfortunate loss of opportunities to gain economically has indeed been due to presence of oil, the major resource curse. The more innovative aspect of the analysis of the resource curse in these Muslim majority countries is the attribution of poor governance and institutional failure as key factors causing the severe lag in economic performance.

The other innovation in the book is the study of the resource curse from an Islamic lens. Would application of the fundamental Islamic principles of justice and equal distribution of resource benefits been able to reverse the resource curse? This book provides an illuminating understanding of the Qur'an that the gatekeepers of countries endowed with resources are duty bound to exercise the highest standards of morality and good governance to ensure fair distribution of the benefits from the resources gift from God to the population. This analysis of the impact of poor governance on managing resources is timely at this time when there is a global momentum to put in place good governance across the whole spectrum of political, social, financial and economic management. This book makes invaluable contributions on the merits of applying Islamic principles on development and distribution of wealth emanating from resources.

The authors provide policy recommendations that are aligned within an Islamic framework and discuss at length how these Muslim countries can achieve the ideal structure envisioned by the Qur'an. They trace the fundamental differences between Institutions from the conventional and the Islamic perspectives and present why an Islamic institutional framework offers the most complete solution for the countries managing rich resources.

This added Islamic perspective on managing resources makes recommendations on resolving the resource curse as one of the most comprehensive in resource curse literature. These recommendations of adopting Islamic principles on economic management as envisioned in the Qur'an are also modern in terms of the emphasis on the importance of building strong institutions and leadership practising good governance as essential in deriving optimal rewards from the resource gift. Islamic principles of management ranging from rules of behavior relating to property rights, trust on contractual arrangements and conduct of economic dealings to economic policies undertaken with social justice and high standards of morality, are all aligned with the current global concerns of promoting social unity in the market place for more equitable distribution of wealth in the pursuit of social justice.

There is no evidence that prior research had looked at Islamic teachings to ameliorate the resource curse problem. This book is an important contribution and could serve as a reference in Islamic economics and studies to manage depletable resources.

The country specific studies on Malaysia, Saudi Arabia and Qatar allow readers to understand the extent of the resource curse issue in leading resource rich Muslim countries. They provide the stark reality that corrosion of quality of institutions in these countries and the misallocation of the vast wealth from natural resources have resulted in imbalanced and unsatisfactory economic conditions. The authors provide sound justification on the need to change the current institutional framework.

In the case of Malaysia in particular which made great strides in affirmative action policies for more equitable distribution of wealth, a more Islamic slant in using proceeds of resource development could have avoided the channeling of benefits to certain groups and the income disparity in the Muslim population may not remain so stark today, more than 40 years after implementing the New Economic Policy. The Malaysian country study demonstrates the more acute impact of political actions undermining competitive behavior to favor select groups. Non-adherence to Islamic principles of distribution of the resource wealth show how the behavior of the political elite entrusted to take care of it citizens can actually deepen the resources curse. Policies which aim to maintain power while weakening institutions further deepen the social injustice in resource management. Results could have been different for Malaysia if leaders had sustained the institutions built by the forefathers and applied the globally accepted principles of fairness and social justice through good governance in managing the rich resources in the country. The analytical content and evidence based analysis make this book a valuable contribution to the understanding of the resource curse problem in Muslim countries. The suggestions in adopting specific principles of economic management aligned to fundamental Islamic principles of justice in distribution of resource benefits are worth serious consideration for resource rich Muslim countries to begin their path towards building strong institutions and practicing good governance.

PREFACE

There is considerable evidence to support a negative link between natural resource endowment and economic development and growth, which is commonly referred to as the 'resource curse' (Yuxiang and Chen, 2011). The membership of the Islamic Cooperation Countries (OIC) is composed of 57 countries (that include Palestine), several of which are rich in oil and natural gas. Given significant oil and gas exports revenues, one might expect that a number of the OIC countries would be at the forefront of economic performance and growth as they develop their 'manna' from heaven, especially during times of historically high oil prices. Yet, the performance of the oil-producing OIC countries has been inferior to the economic performance of comparable non-resource-rich countries.

During the period of rising oil prices, the economic performance of oil-producing OIC countries lagged behind that of similar countries that were not resource rich. Poor economic performance suggests the existence of the resource curse, whereby instead of raising wealth, natural resource endowments traps an economy in a lower economic equilibrium. In the current environment of lower oil prices and with the expectation that oil prices will not recover any time soon, the path to economic stability may be even more arduous with countries having to navigate out-of-budget shortfalls. There has not been a better time to investigate the reasons for this subpar performance and to assess the popular belief that a few have engaged in constant accumulation or manipulation of wealth at the expense of the overall economic benefit and growth for the masses in these countries.

In a study of ten OIC countries, we determine whether they have achieved sustainable growth and superior economic performance. On this premise, we assess whether the symptoms of the resource curse are present and whether natural resource abundance in the OIC countries has been a curse or a blessing. Our goal is to present policy recommendations for the OIC countries to benefit more from their resource wealth by adopting welfare-enhancing institutional and policy reforms in line with the framework envisioned in the Qur'an and practiced by the Prophet Mohammad (sawa) in Medina.

ACKNOWLEDGMENTS

We are indebted to Datuk Latifah Merican Cheong for her generous foreword to our modest contribution. We thank the panel of anonymous referees for their helpful comments and suggestions. The support and dedication of Dr. Koohi-Kamali and her outstanding team to this project are greatly appreciated. Last but not least, we acknowledge the support of our families without which this book would not have been possible.

ABBREVIATIONS

APF	Alaska Permanent Fund
CIA	Central Intelligence Agency
CSF	Copper Stabilization Fund
EITI	Extractive Industries Transparency Initiative
FDI	Foreign Direct Investment
GCC	Gulf Cooperation Council
GDP	Gross Domestic Product
GMM	Generalized Method of Moments
ICRG	International Country Risk Guide
IMF	International Monetary Fund
MG	Mean Group
NDP	National Development Policy
NEP	New Economic Policy
NNP	Net National Product
OECD	Organization for Economic Cooperation and Development
OFAG	Oil Fund for All Generations
OIC	The Membership of the Islamic Cooperation Countries
PMG	Pooled Mean Group
SPF	State Petroleum Fund

SWF	Sovereign Wealth Fund
UAE	United Arab Emirates
UMNO	United Malay National Organization
WDI	World Development Indicators

· 1 ·

THE OIC COUNTRIES

Introduction

The Organization of the Islamic Cooperation (OIC) is the second largest international organization after the United Nations, with a membership of 57 countries (that includes Palestine) and spanning four continents. The member countries are mainly classified as developing countries and account for more than one fifth of the world population.

The OIC has gained status and has tried to project Islamic solidarity in the contemporary world. The OIC has endeavored to distance itself from inter-Muslim rivalries, but unfortunately in recent years it has taken sides in favor of influential majorities. According to Rehman and Askari (2007) OIC membership is typically based on the following; (1) governments who have adopted Islam as the official state religion, (2) Islam is practiced as a primary religion, (3) the country has a significant Muslim population or (4) self-declaration as an Islamic republic. Although these countries have different political, economic and socio-cultural backgrounds, they represent the voice of the Muslim world and espouse the importance of safeguarding the interests of Muslims in the world in the spirit of promoting international peace and harmony.[1]

1. About OIC-OIC.org retrieved on 2016–02–29

The OIC includes among others, the following key objectives in its charter; (1) to enhance solidarity and fraternity among member states, (2) to safeguard and protect the common interests and support legitimate causes of the member states, (3) to promote inter-state relations based on justice, mutual respect to ensure global peace, security and harmony, (4) to exert efforts to achieve sustainable and comprehensive human development and economic well-being in member states, (5) to reaffirm the rights of people as stated in the United Nations Charter and international Law and (6) to promote, preserve and disseminate Islamic teachings and values based on moderation and tolerance as well as promoting Islamic culture and safeguarding the Islamic heritage.

OIC Countries and the Resource Curse

The OIC countries have high economic potential in various fields, including productive land, mining, energy, agriculture, human resources and financial means. Hassan (2009) categorically classifies the OIC member countries into 3 sub-groups—the Least Developed, Middle Income, and Oil Exporting to illustrate the different levels of economic development and endowment within the membership.[2]

Collectively, the member countries of the OIC contribute over 9% of the world's total GDP[3] with 22.8% of the world's population. Saudi Arabia and the United Arab Emirates alone contribute about 17% of world oil production.[4] It could be expected that the wealth of natural resources would position the oil-producing OIC countries at the forefront of economic performance and growth. Yet, their economic performance often falls behind other countries. Askari and Rehman (2013) state that the poverty level in

2. The Least Developed countries include Afghanistan, Bangladesh, Benin, Burkina Faso, Chad, Comoros, Djibouti, Gambia, Guinea-Bissau, Maldives, Mali, Mauritania, Mozambique, Niger, Senegal, Sierra Leone, Somalia, Sudan, Togo, Uganda and Yemen. The Middle-Income countries are Albania, Cameroon, Cote D'Ivoire, Egypt, Guyana, Indonesia, Jordan, Kazakhstan, Kyzg Republic, Lebanon, Malaysia, Morocco, Pakistan, Palestine, Surinam, Syria, Tajikistan, Tunisia, Turkey and Uzbekistan. The Oil Exporters are Algeria, Azerbaijan, Bahrain, Brunei, Gabon, Iran, Iraq, Kuwait, Libya, Nigeria, Oman, Qatar, Saudi Arabia, Turkmenistan, and the United Arab Emirates.
3. CIA- The World Factbook.
4. International Energy Agency.

OIC countries has not only been persistent but has also risen over the last 30 years. The same can be said regarding the state of income disparity for the OIC countries. Whilst income disparity has generally risen in the Organization for Economic Cooperation and Development (OECD) countries over the past 10 years, income disparity in the OIC countries has remained constant over the last 30 years.

The subpar economic growth performance of the OIC countries may be a manifestation of the "resource curse." The resource curse hypothesis indicates a situation where countries are unable to attain satisfactory economic growth despite being endowed with natural wealth. The reasons for the curse continue to be debated and vary from country to country. We believe that the institutional framework and government policies of OIC countries are not able to transform their wealth (capital) underground into productive capital and achieve satisfactory economic growth and reduce poverty.

In principle, OIC countries should be the exemplars of operationalizing fair distribution of wealth through institutional and governance framework defined by the principles of Islam. According to Mirakhor and Askari (2010), a successful Muslim society is built upon the foundation of the rules prescribed by Allah (swt), the Law-Giver. Such rules are monitored where there are rewards for compliance, and sanctions for non-compliance. As each individual complies with such rules, there is less ambiguity with more degree of certainty in the expected conduct of others in the society. In such a rule-compliant society, trust and cooperation are enhanced and transaction costs are reduced, resulting in more investment and growth. Just and well-managed societies should in principle demonstrate just distribution and redistribution of its economic wealth as they allow for continuous, stable growth.

Askari, Iqbal, Krichene, and Mirakhor (2012) suggest more broadly that such a framework promotes, among others, social justice, unity, trust and social cohesiveness. It is also expected that such a framework is able to curb counterproductive behaviors such as greed, deceit, misrepresentation and the misappropriation of property, thus leading to an economy that is flourishing with justice as its hallmark.

The Qur'an provides injunctions that wealth is a blessing of the Creator and is given to provide support for the lives of all mankind and extremes of richness and poverty must be avoided in order to achieve a balanced and just society. There is clear guidance on how to avoid emergence of such extreme imbalances and injustices in the society. Rules prescribed by the Qur'an impose limits on consumption through prohibitions against overspending (*israf*), waste (*itlaf*), and ostentatious and opulent spending (*itraf*). The property rights rules of Islam ordain that the net surplus, after moderate spending to sustain a modest living standard, must be returned to the less able members of the society (Askari et al., 2012). The interest of all generations is to be protected, something that, in turn, has enormous and overriding policy implications for the management of oil and other depletable resources.

Depletable resources are the Almighty's gift to His Creation, which means humanity as a whole. These resources must be transformed to alternative forms of capital to benefit all generations in an equitable fashion. We emphasize that all members of the community, the present and future generations, must receive equal benefits from the transformation of all depletable resources. This requires efficient resource transformation and equal sharing among all members of the community.

Conclusion

The OIC countries should be better positioned to implement the framework according to the principal teachings of the Qur'an. But are the OIC countries respecting and adhering to this complete framework? If not, how far are they from achieving the ideal structure as envisioned by the Qur'an? This discussion leads naturally to the primary motivation for this research. Are oil-producing OIC countries managing their depletable resources within the Islamic framework outlined in the Qur'an? And what do they need to do to be at the forefront of achieving sustainable economic growth and alleviating poverty as required in the Islamic conception of social justice? Is this the case for oil-producing countries or does their economic performance provide evidence to the contrary? If so, could the poor economic performance suggest the existence of the resource curse which instead of raising wealth and natural resource endowment instead traps the economy in a lower economic equilibrium? Does the resource curse explain why extreme wealth and opulence exist alongside widespread poverty in these societies?

Our goal in this study is to provide policy recommendations on how the countries could benefit more from their resource wealth and "cure" themselves of the disease by adopting welfare enhancing reforms in their policies and institution building.

References

Askari, H., Iqbal, Z., Krichene, N., & Mirakhor, A. (2012). *Risk sharing in finance*. Singapore: John Wiley & Sons (Asia).

Askari, H., & Rehman, S. S. (2013). A survey of the economic development of OIC countries. In A. Mirakhor & Z. Iqbal (Eds.), *Economic development and Islamic finance* (pp. 299–324). Washington, DC: World Bank.

Hassan, M. K. (2009). Economic performance of the OIC countries and the prospect of an Islamic common market. M. Kabir Hassan working paper 461, January 2009. *Economic Research Forum*,(p. 4), 461.

Mirakhor, A., & Askari, H. (2010). *Islam and the path to development*. New York, NY: Palgrave Macmillan.

Rehman, S. S., & Askari, H. (2007). An economic islamicity index. *Global Economy Journal*, *7*(2), 3–4.

· 2 ·

THE RESOURCE CURSE—THEORY, EXPLANATIONS AND ITS REVERSAL

Introduction

The resource curse phenomenon suggests that the presence of abundant natural resources may hinder economic growth. According to Cavalcanti, Mohaddes, and Raissi (2011) one of the influential works in the resource curse study was conducted by Sachs and Warner (1995). Sachs and Warner demonstrate that after controlling for a wide variety of variables, an increase of one standard deviation in natural resource intensity would lead to a reduction of about 1% per year in economic growth (Bulte, Damania, & Deacon, 2005). The study has subsequently inspired a large volume of empirical research specifically in determining the various causes of the resource curse. Various factors such as political instability and conflict (including armed conflict) have been identified as common causes.

Many academics have provided significant contributions in their attempt to decipher the resource curse puzzle. The vast literature on the resource curse over the past decade shows how research and work on the subject have evolved. At the inception of the resource curse notion, published materials focused on both exploring and validating the theory. In these studies, while the countries and the resource sectors under analysis may have varied, the focus of the research was similar. Could higher dependence on a depletable natural resource be directly associated with lower economic growth? As the theory on the resource curse gained traction and momentum, the second wave of work saw researchers undertaking empirical studies on a number of different variables to further determine the reasons and causes for the curse.

To the best of our knowledge the term "resource curse" was coined by the Saudi executive director at an executive board meeting of the International Monetary Fund (IMF) in 1980/1981 in an exchange with the representative of Mexico who felt that oil reserves were an unquestioned blessing. We now follow the conventional explanation for the resource curse and begin with the prominent work of Sachs and Warner.

The Resource Curse Theory—Sachs and Warner

The notion that natural resources are a curse rather than a blessing was promoted by Sachs and Warner (1995, 1997, 2001). Since then, this puzzle has attracted considerable interest in development economics and natural resource economics research (Hodler, 2006; Mideksa, 2013). In their work, Sachs and Warner assert that countries with significant natural resource wealth tend to grow more slowly than resource-poor countries. They are of the view that while empirical support of the resource curse is not indisputable, the results strongly indicate the absence of a positive correlation between natural wealth and economic wealth (growth).

From one of their observations, the authors provided the view that resource-rich countries such as oil-exporting countries in the Persian Gulf, Nigeria, or Mexico and Venezuela have not experienced sustained rapid economic growth. By utilizing the repeated regressions method and using growth data from the post-war period, the authors found that high resource intensity tends to be associated with slow growth.

Sachs and Warner mention that none of the countries with extremely abundant natural resources in 1970 grew rapidly for the next 20 years, even after using a variety of different measures for resource abundance (Sachs & Warner, 2001). The authors add that most countries that did grow rapidly during this period started as resource poor, not resource rich. Further details on the empirical findings presented by Sachs and Warner are as follows:

1. There is evidence of an inverse association between natural resource abundance and growth during the period 1970–1990 even after

controlling selected regressors, which include global commodity price shocks, government expenditure ratios, investment rates and long-run changes in the country's external terms of trade.
2. Asian countries that are resource-poor, such as Singapore, Hong Kong and Korea have grown faster than resource-rich Malaysia and Thailand.
3. In a separate sample from the same study of seven oil economies namely Bahrain, Iraq, Kuwait, Oman, Qatar, Saudi Arabia and United Arab Emirates, there is an even stronger negative association between growth and natural resource abundance as indicated by the countries' slow economic growth.
4. Although there are 23 developing economies that are considered in the top quartile of resource-abundant countries, the striking evidence suggests that only two countries have experienced sustained high levels of economic growth. These countries are Malaysia and Mauritius, with recorded per capita growth greater than or equal to 2.0 percent per annum. However, in these cases, the authors indicate that manufacturing exports, rather than resource-led growth, account for the sustained high levels of economic growth.[1]

The work of Sachs and Warner also explored possible explanations for the resource curse. The first explanation is what is commonly referred to as the Dutch Disease, a condition where the existence of large natural resource sectors or booms in these sectors affects the real exchange rate, which in turn increases the profitability of producing non-tradeables, such as housing, versus tradeables that include non-resource exports, and thus impacts the distribution of employment throughout the economy (Sachs & Warner, 1997). In the Sachs and Warner version of the Dutch Disease model, there are three sectors in the economy as seen in Table 2.1:

---

1. Countries studied: Algeria, Austria, Argentina, Bangladesh, Belgium, Bolivia, Botswana, Brazil, Cameroon, Canada, Chile, China, Colombia, Congo, Costa Rica, Cyprus, Denmark, Dominican Republic, Ecuador, Egypt, Fiji, Finland, Gambia, Gabon, Ghana, Greece, Hong Kong, Iceland, India, Indonesia, Iran, Ireland, Israel, Italy, Ivory Coast, Jamaica, Japan, Jordan, Korea, Kenya, Malawi, Madagascar, Malaysia, Mauritius, Mauritania, Mexico, Morocco, Netherland, New Zealand, Nicaragua, Nigeria, Norway, Pakistan, Panama, Paraguay, Peru, Philippines, Portugal, Rwanda, Senegal, Sierra Leone, Singapore, Syria, Spain, Sri Lanka, Sweden, Switzerland, Taiwan, Thailand, Tunisia, Trinidad & Tobago, Uganda, Uruguay, USA, Venezuela, Zambia and Zimbabwe.

Table 2.1: Sachs and Warner Dutch Disease Sector Breakdown.

No.	Sectors	Remarks
1.	Tradeable Natural Resource Sector	Capital and labor are not used.
2.	Tradeable Manufacturing Sector	Non-resource sector. Capital and labor are used.
3.	Non-Traded Sector	Capital and labor are used.

According to the model, the greater the natural resource endowment, the higher is the demand for non-tradeable goods and consequently the smaller the allocation of labor and capital to the manufacturing sector (Sachs & Warner, 1997). The authors observed that when natural resources are in abundance, the concentration of tradeable production would be in natural resources rather than in non-resource-based sectors such as manufacturing. There is a tendency for capital and labor that would have been employed in manufacturing to be pulled into the non-traded goods sector. Consequently, when an economy experiences a resource boom and the real exchange appreciates the tendency is for the manufacturing sector to shrink and the non-traded goods sector to expand. In addition to the Dutch Disease theory, the authors also explored the political economy of significant natural resource endowment by looking into how natural resource production could generate high economic rents. They presented conditions where governments typically earned most of the rents from natural resource exploitation and where there were impediments to innovation erected by special interest groups, as these groups are more powerful when they can obtain government revenues from easily taxed natural resources.

In determining whether legal and government institutions are efficient, the authors utilised the rule of law index[2] and their empirical results indicate a positive association between the quality of legal and government institutions and economic growth. Finally, the authors also saw that natural resource abundance may affect growth directly through the degree of trade openness. Higher primary exports tend to promote economic closure and lower openness to trade. The work of Sachs and Warner has grown to become one of the most prominent in the resource curse theory (Al Rawashdeh & Maxwell, 2013; James & Aadland, 2011).

2. This index is provided by the Political Risk Services, a company that also produces an index for bureaucratic quality (Kolstad & Wiig, 2012).

Subsequent literature supporting the resource curse theory followed Sachs and Warner's empirical specification and introduced new explanatory variables for resource dependence and abundance, while others derived theoretical models that were loosely related to their empirical specification (see Bjorvatn, Farzanegan, & Schneider, 2012; Hodler, 2006; Mideksa, 2013; Williams, 2011). Some of this literature is discussed in the following paragraphs.

Supporters of the Resource Curse Theory

The resource curse and the paradox of natural resource wealth have inspired numerous studies and continue to fascinate academics across disciplinary boundaries (Cockx & Francken, 2014). According to Cockx and Francken (2014), it has become almost conventional wisdom to observe and note that natural resources do not necessarily generate economic prosperity as they represent a curse rather than a blessing. A model developed by Hodler (2006) attempts to explain why natural resources are a blessing for some countries but a curse for others. The research highlights natural resources as the cause for fighting between rival groups, such as those that exist due to ethnic fractionalisation. As a result, productive activities are reduced and property rights weakened.

In support of this argument, Hodler provides empirical evidence to show that an oil-rich country such as Norway performs well because of its homogeneity as compared to countries such as Nigeria. This led Hodler to point to ethnicity as a fundamental source of the curse. Along similar lines, a study by Larsen (2005) on the effects of natural resources and growth across different countries also indicate most resource-rich countries across the board experience the resource curse. Auty (2007) acknowledges the effect of complacency and neglect on managing natural capital and explores possible solutions to combat the curse, such as reinforcing sound management of natural resources and providing an index of policy sustainability in the form of the net saving rate. He suggests that by the policy index and along with other measures such as a capital fund for sterilizing economic rent, initiatives to increase the transparency of rent flows and the rigorous evaluation of alternative uses of additional public sector revenue can improve the efficiency by which natural resource rent is transformed into alternative forms of capital to sustain rising social welfare.

Askari (1990) show that oil revenues should not be consumed because if oil revenues are used to support consumption, then Net National Product (NNP) and consumption will drop sharply when oil resources are depleted. Depending on a number of factors, savings must be increased to compensate for oil depletion so that consumption over time is smoothed. In order to ensure reasonable savings from oil resources and to account for the future when oil is depleted, a country should start to save an extra amount as it begins to deplete its oil resources. The summarized equation, or policy prescription, for an economy totally dependent on oil, the required savings rate is given by:

$$\hat{s} = 1 - rT\,(1 - s)$$

Where: $\hat{s}$ is the required savings rate, T represents the number of years till oil runs out, s is the desired savings rate when oil runs out and r is the real rate of return on investment. Accordingly, if r = 1 T = 40, s = 20%, the conventionally measured savings rate, $\hat{s}$, now should be 68%. In a nutshell, if the economy (for an economy that is totally oil based) has a utility function consistent with eventually having a savings rate of 20% (after oil has run out), the necessary policy is to now invest approximately 70% of the national product in productive capital (Askari, 1990). Of course, this required savings rate would be lower to the extent that the country has a non-oil NNP and higher in the rate of return on investment. The latter indicates that to the extent the country chooses good investments (with effective institutions and thoughtful policies), it does not have to save as much today.[3] This framework thus provides the basis that poor management of resources by the government today severely damages a country's ability to ensure sustainable income and commensurate benefits for future generations.

Kolstad and Wiig (2012) note that natural resources such as oil seem to stir up counterproductive behaviors on the overall management of the country's resources and growth performance. Their paper tests whether there is a direct effect of natural resources and trust. They conclude that the effect of natural resources on trust depends on the initial (when the natural resource is discovered) corruption level of a country (namely, effective institutions including the rule of law). Latif, Farooq, Loganathan, and Shahbaz (2014) who assert that the resource curse hypothesis has become an important area

3. This requires a simple adjustment, see Askari (2006).

of research for the development economists of this era, investigate the relationship between natural resource abundance and economic growth for the Venezuelan economy. Their results confirm that variables related to natural resource abundance impede economic growth, while financial development, capital stock and trade openness enhance growth.

In another country specific study, Ayelazuno (2014) provides the opinion that capitalistic exploitation and its negative structural effects on the economy cannot be ignored. He formulated a critical political economy approach to study the extent of the curse in Ghana. In his study, Ayelazuno found that the danger of a resource curse epidemic in the country is very real due to the continuous exploitation by global capital and dependence on the export of low-value commodities.

In another study, Askari (1990) found that as a result of inefficient governmental policies, Saudi Arabia faced a grim future largely as a result of inefficient subsidies, low yielding investments, excessive military expenditures and corruption. Gradually research into the resource curse has expanded to exploring the explanation of why it happens, and what could be done to reverse the condition and its negative fallout.

Resource Curse Explanation—Dutch Disease

In determining the economic explanation for the resource curse, a number of earlier studies concur with the Dutch Disease explanation promoted by Sachs and Warner. Algieri (2011) and Dülger, Lopcu, Burgaç, and Balli (2013) corroborate its presence in a study examining the Russian economy. Algieri (2011) detects a significant appreciation of the real exchange rate resulting from booming economy pushed by higher oil prices, relative de-industrialization and a reduction in exports from the non-booming sectors. Dülger et al. (2013) find similar effects by employing updated data sets and after accounting for the endogenously determined structural change from past studies.

Beine, Bos, and Coulombe (2012) note that both the Dutch Disease and the strength of United States currency contributed to the performance of the Canadian economy and Kutan and Wyzan (2005) found significant effects of the Dutch Disease in Kazakhstan. Pegg (2010) provides mixed or a nuanced view on the experience of Botswana; in a case study, he indicates that the country suffers from many of the symptoms of the Dutch Disease but not due to the usual causal reasons as posited in the model.

Resource Curse Explanation—The Role of Institutions (Nigerian Disease)

Studies of the resource curse from the Nigerian perspective started to emerge as researchers moved from focusing on the implications of growth and economic performance to extending it to other variables that include government priorities, accountability and strength of societal institutions. Apergis and Payne (2014) espouses that it is improvement in institutional quality over time which will play a significant role in mitigating the downside effects of an oil curse. A number of authors (Everhart, 2010; Kolstad & Søreide, 2009) have looked into corruption and Bjorvatn et al. (2012) found that it is the strength of the government in its ability to rein in corruption that becomes the precursor to growth for resource-rich countries.

In another study, Williams (2011) explores the effects of natural resources, transparency and growth. A negative causal association was found from resource export revenues to transparency leading to a subsequent decrease in economic growth. His study also suggests that institutional explanations lies at the heart of the resource curse problem and explores the effect of levels of transparency on economic growth. Kolstad and Wiig (2012) find that the large revenues of resource-rich countries from resource exploitation invariably lead to institutional degradation, affecting increased patronage and other rent-seeking activities to benefit from resource rents, which subsequently leads to economic inefficiencies.

According to the summary prepared by El Anshasy and Katsaiti (2013), there seems to be a strong association between abundant natural resource endowment, corruption and bad economic performance. They find in a panel data on 79 resource and non-resource countries that it is the quality of fiscal policy and not government size that matters to growth in the group of resource-rich countries. Other theorists postulate that it is institutional quality that determines the presence or absence of the curse (Costa & Santos, 2013).

For example, Boschini, Pettersson, and Roine (2013) categorically group tradeable products into "appropriability". Agricultural products are deemed to have low appropriability in contrast to diamonds or oil. The importance of institutions is related to the level of appropriability, the more appropriable is a resource the more important the reaction with institutions. The authors conclude that resource abundance is turned into an asset rather than a curse when there is sufficient improvement in institutional quality. Much of these results can be deduced from the simple observation that countries that had effective institutions when significant oil resources were discovered (such as Norway, Canada, and the UK) appear to have benefited from the discovery, while countries with ineffective institutions (such as Iran, Nigeria, Saudi Arabia, and Venezuela) had subpar economic performance.

Moreover, it would appear that in the presence of ineffective institutions the discovery of significant oil resources has impeded the development of more effective institutions. Simply said, those in power and their supporters and cronies have no incentive to improve institutional quality and effectiveness. Most clearly, an enhancement in the rule of law, monitoring of corruption and effective enforcement would impede their ability to amass vast fortunes through rent-seeking activities at the expense of economic growth and the general welfare of society. Moreover, there are arguably even more destructive political repercussions. Namely, those in power and their cronies are motivated to hijack the political system and pass on their rent-seeking activities to their offspring and associates. Political, social and economic reforms become impeded under such conditions. Thus, the vicious cycle is fueled by resource depletion and little hope for a turnaround until a benevolent ruler emerges or the resource is depleted.

Mehlum, Moene, and Torvik (2006) show evidence that resource abundance provides a positive effect on growth in the presence of high quality institutions and thus the negative effects of the curse may be addressed through institutional strengthening. There is a significant negative relationship between resource abundance and growth in their data sample and a significant positive relationship between resource abundance and institutions. This implies that resource abundance reduces growth to a smaller extent if a country's institutions are effective (Mehlum et al., 2006) and ultimately enhances growth if the institutional structure is of high quality.

Along similar lines, Bulte et al. (2005) explain the importance of institutions in explaining the role of natural resources on growth. The authors examine whether the paradoxical result that resource-rich countries have lower economic growth than resource poor countries applies to the relationship between resource wealth and measures of human development. In contrast to prior studies, the authors considered a broader set of welfare and development criteria. Their results indicate the presence of resource curse in economic growth. In addition, the authors were able to identify a spillover to a broader set of development indicators. For example, they found countries that rely heavily on natural resources are seen to underperform worse in numerous other areas. Ceteris paribus, these countries also have lower human development indicators and life expectancy; and a higher percentage of the population was found to suffer from undernourishment or lack of access to safe water. All of this would also support the presumption of a highly skewed wealth and income distribution. Secondly the authors provide evidence that all else being equal, countries with abundant natural resources end up with less effective institutions and with governments that perform poorly in several areas.

The authors also found that for countries with population without access to safe water, a one standard deviation improvement in the rule of law or government effectiveness indicators moves a country's development outcome upward by about one-fourth of a standard deviation. Such an improvement amounts to a 4% reduction in the population without access to safe water and an increase in life expectancy of 3.3 years. In summary, Bulte et al. prove that the resource curse phenomenon has a much broader impact than simply on economic growth and that institutional reform is imperative if countries are to develop and grow in most areas. Institutional quality is the foundation of economic performance, whether a country has, or has not, abundant resources at its disposal. Other views suggest that natural resource abundance can result in the deterioration of a country's governance.

For example, Robinson, Torvik, and Verdier (2014) specifically state that natural resource endowments lead to the dysfunctions of the public sector and unsustainable budgetary policies. They study the categories of political incentives that were created as a result of the resource booms and how this in turn led to adverse consequences for national income and development.

Robinson et al. group income from natural resource depletion into two periods: the first is stated as the time when income from natural resources accrues to the government and the incumbent must decide on the amount of resources to be extracted in the first period; and the second period relates to how much will be left for the future. Election is said to take effect at the end of the first period. The authors also suggest that resources are used either via the incumbent consuming it or by distributing them as "patronage" to influence the outcome of elections. They model patronage as the offer of employment in the public sector based on earlier political science literature, which indicated that politicians often distribute public jobs or special favors to gain votes during elections. Thus, there is institutional linkage, which the authors chart as revolving around resource rents and how political advantage is gained through the manipulation of public employment. Robinson et al. explain the following as the implication of this practice of patronage:

1. Politicians, while they are in power, tend to extract natural resources excessively and not according to the extraction path that is socially efficient.
2. Resource booms also increase resource misallocation in the rest of the economy as ruling families, dictators and politicians find ways to increase patronage to stay in power. As a result, public sector employment, which is deemed as being relatively inefficient, increases.
3. Countries that do not build strong institutions to promote accountability and state competence will suffer from the resource curse given the perverse political incentives that are created by a resource boom.

The authors imply further that whilst they agree that inter-sectorial resource allocation is an integral element of the resource curse; their theory is driven by political and not economic incentives. Citing examples provided from earlier literature, the authors provide evidence that a resource boom was directly linked to a political strategy for expanding the public sector. They note that the over-expanded public sector in response to the windfall income of Trinidad and Tobago was attributed to a resource boom triggering a high government share of formal employment. Similar effects were seen in other countries such as Ecuador, Venezuela, Mexico and Zambia. Their hypothesis essentially states that incentives politicians face when they confront resource booms map into different policy choices depending on the quality of institutions. Low quality institutions result in making political strategies infeasible or generally unattractive.

In summary, the results from the study point to inefficiencies and shortcomings in the economy due to politically motivated, wasteful and inefficient redistribution of proceeds from the sale of natural resources as politicians attempt to influence elections in their favor in order to hold onto power. The extent of this phenomenon actually depends on the quality of the institutions. Countries that are effective in limiting the ability of politicians to exercise patronage are able to raise national income from resource booms. In the absence of effective institutions to curb the problem, perverse political incentives would dominate, therefore, leading to a fall in economic growth.

Related ideas have been developed in several other studies. For example, Kolstad and Søreide (2009) propose that priority should be given to policies that address rent-seeking and patronage activities. They note that policy in resource-rich countries should be less about macro-economic management and more about institutions so that the right incentives could be provided to players in the resource sector. According to Kolstad and Søreide, corruption is the main reason why resource-rich countries perform badly in economic terms, as resource rents make rent seeking a profitable practice. Policies that promote transparency, effective monitoring and appropriate punishment to corrupt officials, no matter who they are, would be helpful in transforming the curse into a blessing. As a result of perverse incentives, the time and skills of individuals are spent in seeking ways to exploit resource rents at the expense of engaging in productive practices.

In a similar argument with that of Robinson et al. (2014) above, Kolstad and Søreide note resource revenues induce patronage as governments pay off supporters to stay in power, thus resulting in misallocation of public funds and reduced accountability. In an effort to shed some light on how natural resources management is corrupted, the authors are of the view that three structural preconditions, which include rents, authority and opportunity, have to be met. They elaborate each of these areas as follows:

1. Rent: Rents from natural resources constitute windfall gains and are in the form of foreign-exchange gifts. This type of rent differs from other sectors that would have rents depending on some form of market failure. There is greater potential to reap high rents due to the lack of competition in the industry and oligopolistic behavior of end markets, which also inspire certain agents to appropriate part of the rents to engage in corrupt practices.
2. Authority: Natural resources are typically under state ownership and, therefore, decision making for extraction and distribution is also under their purview. As a result, decisions relating to natural resource exploitation often become national interests or diplomatic concerns.
3. Opportunity: The extraction of resources often requires complex financial, contractual and technology. It is often difficult to establish clean vendor management with transparent and rigid procedures. Opportunities to conceal would be more prevalent in resource extraction as compared to other sectors. The constant business dealings between private and public officials could also provide scope for misuse and exploitation such as in the awarding of resource licenses. Bureaucratic corruption can easily take place given the potential to gain significant profits as well as greater opportunities to conceal such transfers.

Kolstad and Søreide further state that all of the above preconditions are particularly evident in most natural resource economies. Thus, from their perspective, countries that have "bad" institutions suffer a resource curse, while those with "good" institutions do not. Their focus on "good" institutions specifically refers to institutions with politicians that will be held accountable for the use of public resources. The authors note that it is also important to understand the interests and incentives of key players in resource-rich economies to alter incentives and redirect counterproductive behaviors. The authors refer to the direct distribution of resource revenues for a potential solution to the resource curse problem. However, they note its implementation challenges and conclude that the work of finding a complete solution still requires further research.

Acemoglu, Johnson, and Robinson (2005) attribute commitment problems of those holding political power as the reason for the divergence between the interest of the state and citizens. Agents who control the state are typically seen as self-interested and the situation where commitment becomes a problem arises when those who are in power make promises that they are unable to commit to or fulfill. Acemoglu et al. provide the example of a dictator promising to obey the rules of democracy. The promise is, however, not credible due to the lack of a higher authority that ensures that he will fulfill his promise. The monopoly hold on military and political power means that the power to enforce authority is at his disposal. The authors view the essence of the problem as the absence of an impartial party to have a sound will and power to manage and enforce agreements.

Another cause of the problem is the desire of political elites to protect their power. Political power is often regarded as the precondition to protect their livelihood as they may derive their sources of incomes, rents and preferential privileges from it. Policy decisions relating to economic change are made in light of ensuring that the hold on power does not become eroded. Consequently, any reformatory proposals for better institutional changes will be opposed due to fear of losing political power, even if they may serve to benefit society as a whole.

Acemoglu et al. state that this idea can be commonly seen in the literature on international trade where individuals or groups erect impediments or blockages to free trade initiatives, even though it may be deemed socially undesirable to do so. This is due to the fact that those invested in certain sectors, which do not enjoy a comparative advantage, would lose out from free trade. The authors offer the following points to illustrate the working of political power within the economic institutions paradigm:

1. The preference that individuals have over economic institutions is driven by the allocation of resources that the institutions can induce.
2. People's preferences are typically not in agreement, because different economic institutions benefit different groups.
3. The state determines the choices surrounding economic judgments and due to the absence of a third party to enforce the commitments of the state, the problem of commitment is severe in political institutions.
4. Those with the higher political say determine the equilibrium structure of economic institutions. Consequently, the distribution of political power determines the outcomes relating to resource allocation and economic growth.

Acemoglu et al. explain that other external pressures may also result in changes as seen in Japan after the Meiji restoration or in Russia after the Crimean War. The authors conclude that only in the light of democratization would there be a better distribution of political power. The social conflicts that arise as a result of their actions cause significant distributional consequences of which more attention is required on their impact. Based on the authors' view, the distribution of political power is crucial in defining and distinguishing between good and bad economic institutions. The strong desire for political power is, therefore, a real concern as it is in the hands of these individuals that the responsibility of distribution and redistribution of resources rests. Decisions made on the basis of self-interest for the gain of pure political power can cause only distorted outcomes for society.

To summarize, within the institutions and resource curse context, the preceding discussions show that the resource curse is often identified as a problem relating to the behavior of economic or political agents. The prescription to ameliorate the curse commonly leads to addressing institutional weaknesses.

Military Expenditures, Armed Conflicts and the Resource Curse Theory

Oil revenues provide rulers the means to build up their military with high military expenditures and arms imports to prevent domestic conflict, deter local opposition and takeover and to buy foreign support from the great powers through arms imports. In some cases, a military build-up may be motivated by the presence of valuable resources in a country next door (such as Iraq's invasion of Kuwait) that could be captured by military means. While military expenditures are a waste of productive resources that do little for economic development and growth, armed conflict and wars can cause unimaginable economic damage.[4]

Le Billon (2001) traces the link between natural resources and war from the 1990s, towards the end of the Cold War. In his view, it was during this period that many armed groups relied on revenues from natural resources to shape strategies of power and commercialization of armed conflict. The end of the cold war resulted in a drop in foreign assistance for many governments, thereby inducing a rush towards mobilizing private resources to sustain military and political activities. According to Le Billon, the control and exploitation of resources motivate and finance parties to engage in conflict. Natural resources increase the vulnerability of countries to armed conflict and undermine political institutions from conducting peaceful negotiations. In support of the resource curse theory, Le Billon suggests that resource abundance has detrimental effects on economic growth and political stability. He argues that abundant natural resources cause both a vulnerability and risk of conflict with their availability a "predictor indicator of conflict." In assessing the factors that specifically are in play in the reproduction and transformation of resource-linked conflicts Le Billon provides the following:

1. The distortionary effects of dependence upon valuable resource on societies can be strongly linked to resources and armed conflicts.
2. Lootability of resources is crucial with regard to the opportunities for interested parties to seize or retain control over resource revenues.

In other words, the motivation and behavior of political parties in designing conflict and war strategies are all centered upon the race to secure access to resources. Le Billon explains that resources are seen as the "prize" that comes with gaining territorial control because it provides armed groups with the "loot" necessary to purchase military equipment, which consequently leads to the commercialization of armed conflict.

Additionally, he highlights that when economic and political control is dominated by a specific group, dissent amongst the ruled increases and those marginalized often see violence as the main route to gain power and to affect

4. Askari (2012).

political change. The degeneration of political systems largely determines the conflicts over resource exploitation and becomes the prime determinant to influence political systems and economic accumulation. So apparent is the violence surrounding resources that Le Billon highlights a distinguished pattern that exists between resources and the type of violence inflicted to gain control over them. He explains that the nature of violence can change and is dependent on whether resources are produced or extracted. In the case of the former for resources such as crops, violence takes a more structural form and is coercive to labor. In some cases, controls over the produced resources are manipulated over trade activities. For the latter, violence is more physical in order to achieve territorial or state control. In addition, he provides that the motivation and funding of conflict are facilitated in part from the fact that extracted resources are often easily accessible to government and rebels because the location of these resources is fixed and commonly known. Often, extraction cannot be relocated and conflicts are mobilized to locations where resources and transport routes are located. When government or rebels fight, it is to establish permanent strongholds over the strategically designated geographical location of resources. In other words, armed conflicts areas are tied to the circumscription of resource locations.

Humphreys (2005) compiled the following six causal mechanisms for the onset of civil war:

1. The "greedy rebels" mechanism—lootability of resources motivates rebel groups to engage in continuous fights over capturing the state. Rebels engage in criminal activity typically when concentration of resources in a particular region of a country leads to the belief among dissatisfied groups that there is much to gain from a seceding state.
2. The greedy outsider mechanism—the involvement of third parties and outsiders from neighboring countries, such as states and corporations who stand to benefit in civil conflicts. For example, Humphrey reported that the secessionist bid in Katanga in Congo was in part, supported and instigated by a Belgian firm Union Miniere du Haut Katanga.
3. The grievance mechanism—the indirect consequence of resource wealth, such as transitory inequality from development process and trade shocks, cause instability and dissatisfaction amongst the groups that experiences these effects. The perceived deprivation of resources and uneven distribution of revenues create "grievances" and trigger violent uprising.

4. The feasibility mechanism—the motivation and opportunity for rebels to finance their cause even if the group was originally formed for other reasons. Exploitation of resources increases the prospects of success in pushing forward their agenda and achieving their goals.
5. The weak state mechanism—natural resource-dependent economies typically create weaker state structures for two main reasons. The first holds that resource abundance generates government revenues that are not derived from the taxing of citizens, thereby shrinking the power that citizens have over them. Government activities are less transparent and incentives to monitor them are weaker with little compulsion to adhere to the demands of citizens. Accordingly, citizens have fewer mechanisms to withdraw support from governments. Second, from the government perspective reliance on natural resources revenues rather than taxation has weakened incentives to develop strong state institutions.
6. The sparse network mechanism—a rentier economy is less conducive in achieving trading partners both internally and between states internationally. Daily economic activity affects how citizens interact with each other and natural resource-dependent economies usually generate low levels of internal trade. Related arguments claim that trade arrangements often achieve peace and stability, as it is often mutually beneficial to adhere to the terms of exchange rather than to break them. It also is associated with greater levels of social cohesion between trading partners.

All of these mechanisms provide an insight on how civil wars are generated based on the underlying relationship between natural resources and conflict. The appropriate response to resolving them requires an approach that is based on the workings of these different mechanisms (Humphreys, 2005).

In another study, Basedau and Lay (2009) argued that natural resources, particularly oil, encourage civil war and are said to provide both finances and motive for armed conflict. However, the authors' work offers a differentiation on a country's level of dependence on its natural resources, as they postulate that it is dependence that makes conflict more likely in the first place. The authors view Humphrey's "grievance" and "sparse networks" mechanism as directly relating to dependence. They conduct empirical tests and prove the importance in distinguishing between resource dependence and wealth, as the former creates a problem and latter is useful to maintain peace. In their view, resource wealth is generally used to maintain internal peace "by combining a huge security apparatus with generous distributional policies." As maintaining such policies is highly costly, countries that succeed in applying them and maintaining peace are those with a high threshold per capita of wealth.

Ross (2004) goes even further into the literature about natural resources, civil war and conflict and reviews 14 cross-national economic studies that cast light on this subject. He asserts that the weight of evidence from the literature does suggest that oil dependence causes conflict but is not enough to determine the duration of conflicts. Oil dependence seems to be more strongly associated with separatist conflicts. Ross also found that most studies indicate that gemstones, opium, or cocoa do not seem to initiate conflict but are seen to lengthen pre-existing wars. While Ross does not discount the fact that natural resources ignite or prolong conflicts, he finds that the discussion surrounding the specific qualities and triggering mechanism remains unclear. Ross also states conflict often happens for resource-dependent countries because of the resource curse.

No matter the cause, the literature on natural resource abundance, military buildup and resulting armed conflict and war is unanimous in how conflicts tend to be highly dysfunctional to societies. When we look at the trend in armed conflicts by region between 1946 and 2009, the Middle East ranks high on any scale (UCDP, 2010). The peak in European conflicts came in the early 1990s and the peak in American conflicts in the 1980s. In the Middle East, the number of conflicts increased in the 1980s and has been generally constant over time (again, excluding figures for the protests and conflicts that erupted in the spring of 2011). However, there has been an upward trend in the number of conflicts in Asia and Africa. Asian and African conflicts account for a significant portion of the total. Some plausible reasons are that these are populous countries with a larger proportion of younger and low-income groups, and some may have low economic growth potential (little hope for a better future) with unstable and oppressive governments.

Since 1980, the member countries of the Gulf Cooperation Council (GCC) have been among the highest importers of arms in the world. But even more devastating for economic progress have been the wars involving the major oil-exporting countries in the Persian Gulf and their fallout. The three wars that stand out are the Iran-Iraq War (1980–1988), the First Gulf War or Operation Desert Shield (1990–1991), and the Second Gulf War or the Invasion of Iraq (2003). In our opinion these three wars have their origins in natural resources and are an integral part of the curse. The first of these wars was motivated by enhanced access for Iraq to the Persian Gulf and the possibility of grabbing Iran's vast oil resources in Khuzestan (the Iranian province bordering the Persian Gulf and Iraq). In invading Kuwait (the First Gulf War) Saddam Hussain saw the easy opportunity to acquire Kuwait's vast oil wealth and in turn ameliorate the fiscal pressures facing Iraq. In our opinion the Second Gulf War would never have occurred if Iraq was not seen as a threat to the United States' oil-rich allies in the GCC and if Iraq did not have vast oil resources of its own. Moreover, unresolved conflicts—between Iran and the GCC and among the countries of the GCC—in the Persian Gulf all center on oil and gas reserves.

The cost of these conflicts should be noted:

In the Iran-Iraq War, almost two-thirds of the $1.4 trillion global cost of war was borne by the direct belligerents.[5] Prior to the war, Iran's oil revenues were roughly $30 billion in 1988 $U.S., while Iraq's oil revenues were around $37 billion in 1988 $U.S. The total cost of the war to Iran was equivalent to almost 19 years of Iran's oil export revenues. For Iraq, its burden represented 13 years of its pre-war oil revenues. Iran's cumulative GDP between 1980 and 1988 was $739 billion in constant 1988 $U.S. Thus, the total damage to Iran's economy during the war was equal to about 77% of Iran's cumulative economic output during the war years. Iraq's aggregate output between 1980 and 1988 was $363 billion in constant 1988 $U.S. Thus, its total war-related cost was equal to about 136% of its cumulative economic output during the same period.

In the First Gulf War, the Iraqi and Kuwaiti economies suffered the most damage. Given that Iraq's oil revenues before the invasion of Kuwait were $15 billion, Iraq would have needed almost 18 years of its pre-war oil revenues to pay for the total damage inflicted on its economy. On the other side of the conflict, Kuwait suffered at least $130 billion in budgetary and macroeconomic losses during the invasion and occupation by Iraq. Kuwait also needed 13 years of its pre-war oil revenues to cover the budgetary and macroeconomic damage to its economy.

The invasion of Iraq in 2013 inflicted severe budgetary pain on the United States. More than half of the total estimate for the global cost of the war was and will be incurred by the United States—a figure exceeding $1.7 trillion. But the human cost of the war for Iraq from fatalities, injuries, and displaced population has already exceeded hundreds of billions of dollars according to very rough and conservative estimates.

5. See Askari (2012), Chapter 5.

Apart from the total losses to the direct belligerents, the most profound effect of the three wars on the region may be found in the composition of government expenditures in the Persian Gulf. According to World Development Indicators (WDI) data, in the period between 1989 and 1995, Bahrain's armed forces quadrupled, Saudi Arabia's more than doubled, and the United Arab Emirates (UAE)'s, Qatar's, and Oman's increased by 50% to 75%. The number of armed forces in other countries—except Iraq and Jordan—either increased or did not change significantly.

The result of this far-reaching trend of militarism in the region is the high share of military expenditures in almost all countries in the region. In all of them—except in Iran, where the quality of data, especially for military expenditures, is highly disputable—the share of military expenditures in total government expenditures, is either higher than or equal to the share of public education and health care combined during the 1990s and 2000s. These are heavy fiscal fallouts for the countries of the Persian Gulf—three of which, Saudi Arabia, Oman and Bahrain, are included in our empirical study in this book—and are in part an explanation of their economic underperformance.

More recently Watts (2010) discusses oil terrorism in Nigeria and traces how the ugly pairing between oil and blood has defined the country's political landscape. Watts provides that the central part of Nigeria's oil production, the Niger Delta, has become a breeding ground for rampant terrorist activities (hostage-taking, abduction and kidnapping) with the US Department of State closing off the area as the hotspot for "global terrorism." This is an area, according to the research, where the Central Intelligence Agency (CIA) agreed to be a red flag and took it upon themselves to alert the state's higher officials on the possible apocalyptic consequences as a result of the political tension. The crisis looming over the Delta spiraled after an election voting fraud in 2003. Since then, as the rule of law is nonexistent, the area has become ungovernable with violence, and escalating ethnic tensions. Watts further adds that President Obasanjo, who was president from 1999 to 2007, "deployed notoriously corrupt security forces to the Delta," which in turn led to an increase in violence and threats by militants. He states that with the fight and clamor to gain resource control, state and local governments have been rendered helpless by "militant youth-movements, growing insecurity and ugly intra-community, inter-ethnic and state violence."

Adding to the issues is the on-going conflict between the federal state and the "littoral-state" over the sharing of oil revenues, and in turn making the forming of alliances to collaborate and fight against militants even more unattainable. Watt's research stresses the unmistakable linkage between oil, democracy, civil war and economic growth and points to scholarly consensus from Jeffrey Sachs to the IMF economists who all posit a strong dependence between corruption, resource dependency and economic performance. Watts quotes the notable expert in resource wars Michael Klare who saw oil as a dwindling resource and a key strategy one—"that will be increasingly generative of inter-state conflict" and is associated with the "economization of international affairs." While oil is generally presumed to generate rents and as a result lead to the enhancement of military and security budget, Watts traces the deep conflicts and politics defining the economies of violence through governable spaces that emerge from the race to control oil, what he repeatedly calls the oil complex. He marks each governable space by the "differing sorts of rule in which oil complex has contributed directly to a restructuring of pre-existing forms of governance." The rise of new social forces and conflict against traditional power structures are all mainly attributed to oil. Into the governable space and economies of violence mix, Watts outlines three main components for Nigeria:

1. Space of Chieftainship—The point of origin for Nigerian oil production, the Delta, is placed within the Nembe community in Bayelsa State. Due to joint agreements between the Nigerian Petroleum Company, AGIP and shell exploration efforts were focused on these oil fields and they began to produce high quality petroleum. Violence began as a result of intense competition over the control of the oil industry and can be traced back to the late 1980s at the helm of Nembe Council of Chiefs, who had acquired power from the then King, Justice Alagoa Mingi IX, to negotiate oil royalties. Through indirect rule, much of the Delta was marginalized and isolated and through tradition local chiefly powers were invented and a complex structure of kingship and gerontocratic rule began to take place. From the outset, the nation's law was promulgated so that claim over land rights and oil royalties were given to the customary authorities, the king and the subordinate powers such as the Council of Chiefs and Executive Council. It was in 1991 that the Nembe monarch's ineffectiveness in dealing with the oil companies led to the radicalization of the powers to the Council of Chiefs. The council, gained more power and increased in number from 20 to 90 staff as they also began to control negotiations with Shell. As a result various political factions

struggled and competed for office. The council placed themselves as the negotiator between the oil companies and community groups, and it was not long after that the system became manipulated through collusion with community representatives.

At the same time, an uprising of the Nembe community youth was being organized by a former oil company engineer, Mr. Nimi B. P. Barigha-Amage, who fed local youth facts about the oil industry in order to extract concessions from the oil companies. The council Chief and the engineer also colluded in a mutual agreement whereby the Chief would deploy knowledge of military logistics so that shutting down of stations or seizure of equipment can be constructed. The engineer on the other hand would supply members of the youth community information regarding community entitlements.

As a result of the backing of chief and engineer the youth groups progressed to become the intermediary of the oil companies and their liaison officers. Other militant groups emerged within the youth groups with affiliation to different cultural groups resulting in a deep military conflict within the youth community. The environment became rampant with insecurity and lawlessness. When Bariga-Amage gained the position of High Chief of Nembe, he instituted "a rule of terror and chaos far worse than their predecessors" through his own cultural group, *Teme*. Previous royalty rule shifted to warring factions of youth who were in some sort of way working with the oil companies. Inevitably, the environment grew to be rampant with insecurity and lawlessness, disrupting the chieftainship hold.

2. The Space of Indigenous Groups—The Niger Delta consists of complex ethnic structures. It is occupied by the Ibo majority with a practice of ethnic minority exclusion dating back to the 1950s. Minorities under the Ibo rule sought to establish native authorities of their own such as the Ken-Saro-Wiwa and the Movement of the Survival of the Ogoni people (MOSOP). In 1995, Ken-Saro-Wiwa was hanged with several Ogoni members, charged with treason as MOSOP gained mobilized radical militancy. MOSOP drafted an Ogoni Bill of Rights giving autonomous political powers to the Ogoni people along with the right to control and use a fair proportion of Ogoni resources. MOSOP however, failed to establish the nationalist vision it aimed for, as dissent not only grew with non-Ogoni leaders but also within Ogoni subgroups. It represented, in reality an elite group led by the

elite and failed to represent youth and women's movements. The youth group of MOSOP emerged as militant not long after, causing fractures to MOSOP's wish to establish a unified local government that was representative across the board. In this regard, it was the split between ethnic groups, gender and community class groups that also defined the building blocks for claim over territory and oil. MOSOP fell apart, as other indigenous movements gained power. The struggle to establish new local government areas ensued across dispersed ethnic groups who sought to claim centrally controlled oil revenues, even over oil that was beyond their own jurisdiction. Military conflicts continued to escalate between indigenous groups as they each fight to establish a solidary movement under a civic nationalist cause. The indigenous space describes the power struggle between oil and ethnicity, where the two are inextricably linked and contribute to making the delta even more engulfed in political tension.

3. Space of Nationalism—The race to gain indigenous rights in the Delta ultimately also dispensed a race for an oil-nation state causing fragmented and divided political dissent across major groups. Oil is a national resource over which citizenship claims are developed and where gaining control and access over oil revenues becomes a precondition to state creation. From 1996 to the present, the number of local governments grew from 50 to a staggering 700, and the number of state claims from 3 to 36. Thus, Nigeria currently operates in a three-tier government structure consisting of federal, states and local governments. The birth of new political institutions seems never-ending and they became vehicles of corruption. It was the politics of oil that crippled all efforts to achieve a unified nationalist front.

Indeed, Watt's research, paints a vivid picture on how oil has become central in defining the contemporary political and military conflict landscape in Nigeria. He provides a remarkable insight into the intricate workings surrounding the tension between chieftainship, indigenous institutions and race to establish national rule and how they intertwine to form the complex layers behind the country's military and political agendas to control oil.

Successive military dictatorship has in effect plundered oil wealth in the region due to the never-ending fight between ethnic groups to appropriate oil resources. Caselli (2006) believes that it is because natural resources are more easily appropriated by the governing elites that countries with large amount of natural resources become more susceptible to power struggles and forced changes in leadership. He argues that Nigeria's elite does not effectively invest in the country's development because of the probability of losing power to a successful challenger. As a result, the regime in power often impedes human capital formation, infrastructure development and improvements in the rule of law. Caselli reports that the regime leaders are often motivated by self-interests, and efforts to springboard industrialization and development are essentially seen as time wasting given that it will be accrued to future leaders. To study the root cause of problem in Nigeria, Caselli proposes a thorough review of the incentives of the political elite.

A paper by Sala-i-Martin and Subramaniam (2003) confirms Nigeria's dismal performance on every conceivable metric. The authors state that Nigeria's GDP per capita in 1970 was US$1,113 and is estimated to have remained at US$1,084 in 2000, placing it amongst the 15 poorest nations in the world. On poverty and income distribution, Nigeria fares much worse with the share of the population "subsisting on less than US$1 per day increased from close to 36 percent to just under 70 percent." In terms of income distribution, the authors report that in 1970 the top 2% and the bottom 17% of population earned the same total amount of income. In 2000, the top 2% had the same income as the bottom 55%. At the same time, the authors state the obvious, that these economic developments coincided with the discovery of oil in Nigeria. They report the GDP per capita did not change much even though oil revenues per capita increased from US$33 in 1965 to US$325 in 2000. The GDP per capita reported in 2000, remained at the 1965 level, US$245. Thus, the authors conclude the oil revenue of US$350 billion in total did not add any significant benefit to the standard of living and instead caused it to decline.

The state of turmoil in Nigeria today causes much suffering to members of most communities in the country. Constant bloodshed from ruthless killing and hostage taking in order to gain control over oil resources are clear examples of the effects of unleashed resource curse. The call to ameliorate the curse, to lay the foundation in achieving a social construct embedded with justice and fair distribution, becomes central in the case of the highly distressed nation that is Nigeria today and an important element for human survival. The extensive literature linking military conflict and natural resources we have showcased highlights the importance of finding effective solutions to reversing the resource curse problem.

Assessment of the Explanations for the Resource Curse

Bulte et al. (2005) provide a comprehensive review and divide the explanations given in the literature into three main categories: Dutch Disease models, rent-seeking models and institutional explanations. In contrast to the Sachs and Warner studies, Bulte et al. find little empirical support for the Dutch Disease as a principal explanation of the curse and are of the view that Dutch Disease is only applicable if the manufacturing sector is the only engine of growth for a country's economy.

The second category of explanations is centered on rent-seeking activities and how these are destructive in resource-rich countries. The authors explain that since the evidence in support of the resource curse indicates that resources are detrimental to development only in some countries and not in all, the rent-seeking explanation provided by past investigators may be too generic and broad.

The third category of explanations, the most accepted, revolves around the relationship between depletable resources and institutions. The authors stress that the type of resource matters and the form of governance and its policies are the salient institutional features. The authors find resource-rich countries that have point-resources such as oil fields tend to be dominated by factional and predatory oligarchic polities and are governed by governments that promote each of their respective narrow and sectional interests. The authors note that there is ample empirical support for the key role that political and institutional variables play in economic growth and that resource abundance is a major factor shaping the institutional context.

We should add high military expenditures, armed conflicts and wars to this list. High oil revenues (something akin to manna for rulers and governments) permit high military expenditures, a waste of oil revenues. Moreover, the presence of valuable resources invites internal and external conflicts, as seen in the case of conflicts in the Persian Gulf and Nigeria. Conflicts and armed conflicts drain resources away from productive investments and impair national infrastructure and destroy human capital through injures and casualty. Their toll has in some cases (such as for Iraq) amounted to more than the country's total oil revenues in all the preceding years.

Natural Resource Effects on Financial Development

Another strand of the literature on the resource curse has explored the role of abundant resources on financial development. Van der Ploeg (2010) examined the saving pattern of resource-rich countries. He finds that despite advice from the World Bank to save at least the rents from extracting and selling of natural resources, resource-rich economies typically save less than the amount that is required. Van der Ploeg provides an analysis on how many resource-rich countries become poorer despite having abundant resources.

Yuxiang and Chen (2011) find financial development as an important mechanism for long-run growth, and assert that any impact of resource abundance on the financial sector can inevitably influence the pace of economic growth. In making a contribution to this topic, the authors compiled provincial data of China to analyze the relationship between mineral resource abundance and financial development. Their empirical results provide some evidence of a negative link between the two variables. Interestingly, Yuxiang and Chen also describe the linkage between social capital accumulation and explain that it can also affect the pace of financial development since financial contracts are trust intensive. The authors however, do not test this link.

Aside from these studies, there remains limited research to link growth, financial development and the resource curse particularly for OIC countries. We include testing of financial development in our empirical analysis, however, we believe that the institutional quality channel will turn out to be a far more important channel for the appearance of the resource curse and for addressing it.

Reversing the Resource Curse

Several studies have examined potential recommendations to meet the challenges of the resource curse. For example, Stevens and Dietsche (2008) conclude that quantitative analysis on its own may oversimplify policy solutions required to effect institutional change. Instead, they propose exploring real conflict of interests between domestic social groups across different natural resource countries. Additionally, the authors are in support of comparative research on social and political conditions under which natural resource-rich countries are able to escape the resource curse. Specifically, this includes looking into commonalities across different combinations of economic, political and social variables that lead to changes in the structure of domestic politics and underpinning institutions.

Kolstad and Søreide (2009) believe that policy in resource-rich countries should be less about macroeconomic management and more about institutions to prevent rent-seeking and patronage activities, as well as about giving the right incentives to players in the resource sector. Boschini et al. (2013) explore the different dimensions of how institutional quality affects resource-rich economies. Their work centers on institutional initiatives for the reversal of the resource curse. They find that it is the direct involvement of governments in the resource sectors that determines the potential reversal of the resource curse, rather than the government in its role as the enabler of contracts between private parties in society. Corrigan (2014) concludes that at the center of the debate over the occurrence of the resource curse lies the importance of the role of institutions and the need to maintain high quality governance to reverse the curse. It is, therefore, of the utmost importance to maintain strong quality institutions in resource-rich countries.

Costa and Santos (2013) provide a comprehensive review on what experts in the resource curse literature have stated to be the measures to address institutional weaknesses. They note that past researchers such as David and Tilton (2005) have focused on the government's role in ensuring that income generated from resource exports will be directed toward investments that promote development. These authors offer solutions that include providing adequate incentives and clear definition of property rights, an internal political structure to curb counterproductive behavior by economic agents and prevent corruption, waste and rent seeking. They suggest imposing international pressure associated with the availability of loans by organizations such as the IMF. Costa and Santos mention establishing avenues to achieve transparency, and for this strand of idea, they refer to the work of Subramaniam and Sala-i-Martin (2003) whose suggestion revolves mainly on publishing revenue payments for close monitoring of revenue expenditures.

In another strand of the literature, Stevens and Dietsche (2008) argue that typical policy prescriptions are inadequate to effect institutional change. Firstly, in contrast to the suggestion by Glaeser, La Porta, Lopez-De-Silanes, and Shleifer (2004) mentioned above, the authors explain that investing in human development and capacity building will not suffice because the constraints that each individual will face are said to be too engrained in the system. Secondly, proposing for the international community to undertake assessments on resource-rich countries poses national sovereignty issues and provides little guarantee that external certifications will provide better outcomes. A third suggestion proposing for managerial freedom of technocrats to undertake economic reforms is not effective because technocrats are affected by the political and social conditions that shape the incumbent government's interests and developmental perspective. The challenge lies in how these conditions tend to vary across different regimes.

The authors also note that there is limited benefit in proposing the "ideal" institutions without catering for differences in economic, political and social circumstances. In their view, the way forward is to embark on a country-by-country comparative analysis to search for the social and political conditions under which countries with natural resource were able to enjoy economic success. Whilst quantitative work in this area has highlighted the respective political and economic challenges, it has not been suited to capture the changes in interests and correlations that are equated with causation that has led to the prescription of oversimplified policy conclusions (Stevens & Dietsche, 2008). Thus the study that the authors propose is qualitative in nature and takes into consideration the historical context and the different milieu surrounding economic, political and social conditions.

Sandbu (2006), on the other hand, proposes to address institutional weakness by providing an extensive framework on developing a taxable natural wealth account. The purpose of such an account is to convert resource rents into tax revenues and is set up for citizens, based on an eligibility criteria format, and all of the country's revenues from natural resources are distributed on a per capita basis at regular intervals. The author suggests that an independent agency should oversee all distribution efforts and for the funds to be kept separate from the government under a body called the National Wealth Office. At a fixed time interval, the government will then impose a tax on the account on whatever that has been paid in the preceding period. The purpose of this roundabout flow of funds is for the public to feel as though the country's natural resource revenues belong to individual citizens, just like the other types of income which the author notes can only finance government expenditure after being collected as taxes. The engineering of the funds is stated to work via two ways. The first way, which the author refers to as the "softer" way, would allow tax withholding so that individuals in their national wealth accounts would receive the after-tax amount. The "harder" version refers to citizens receiving their entire pre-tax share of that year's revenues from natural resources. At the end of the year, the amount of tax owed would then be reverse transferred from the accounts. Sandbu believes that the harder version would likely create stronger endowment effects than the softer version.

The main idea behind this account is to ensure that government does not have access to resource revenues except by way of taxing the incomes of the population. However, the author cautions that several actions need to be taken to ensure that the funds remain sufficiently sustainable in light of the country facing depletion and volatile commodity prices. For example, a portion would need to be taken out before they are distributed to the citizen accounts so that the funds that they receive are smoothed beforehand and would remain permanent for them. Alternatively, the portion to be taken out can be in the form of taxes that the government levies on the account. By citing evidence to prove that private sector has proved itself better than the public sector at smoothing windfall income, the author takes the position that it is better to rely on private sector advice for matters relating to smoothing policies. In any case, it is stressed that it remains imperative that individuals are given full information of the per capita amount of the adjustment.

Citizens will then have been familiarized by the magnitude and volatile nature of resource revenues, which in turn will motivate and pressure the government to ensure the money it retains, is managed in a proper manner. In terms of who should be eligible to receive the funds, the author interestingly cites the following groups as recipient worthy:

1. Resident citizens with special considerations to be given to long-term foreign students.
2. Adult citizens and not those younger as a precautionary measure to not provide incentives for a high fertility rate that is said to happen often in developing countries.
3. Households and not individuals where those with more children receiving more funds for schooling children. Here it is again mentioned that incentives for fertility choices need to be developed to curb excessive population growth for fear that households will keep on receiving more money at the expense of others if they continue to have more children.

One of the immediate benefits of the account would be citizens' improved understanding of the fiscal nature of a natural resource-based economy since revenues, taxes and smoothing actions become transparent. The author believes that over time, based on the understanding of the inherent volatility of resource revenues, pressures of overspending would be reduced during boom periods. Another benefit can be seen from citizens wishing to know how the government share is spent, since they are well aware of what portion of the fund is captured from the government. Governmental malfeasance, such as patronage and corruption in the oil sector, would be curbed, as its presence would result in lower transfers of funds to the accounts. In the author's opinion, citizen monitoring can be made easier through greater understanding of the monetary effects of the government's conduct.

In determining the exact rate of income that the government should tax from the account, the author leaves the discussion open and mentions that it should be ideally set at a level that achieves the maximum beneficial income effect. It is worthwhile to highlight that the author supplements this discussion with a historical philosophical discussion of income. The proposal by Thomas Paine in *Agrarian Justice* was quoted as a fundamental principle for handling of resource revenues. According to Paine (1995) "since land must originally have been the common property of humanity, today's poor are entitled to their proportionate share of rents on land," an argument that can be extended to resource rents. Supplemented by other research, the author clearly shows his preference for the direct distribution proposal.

The author is also cognizant to some of the setbacks that may occur from implementing the natural wealth account framework. For example, he highlights that individuals, especially the poor, may consume all the rents, leaving nothing for productive investment to generate growth. Another anticipated worry refers to the diminishing incentive for productive work and thereby reducing labor supply. Unfortunately, the answer to this dilemma was not clear cut. Rather, the author points the discussion to determining whether it is individuals or governments that would behave wastefully. The author provides evidence that private agents had a higher propensity to save as compared to governments and cites the success of micro-finance projects as indicators of individuals with higher ability to direct resources to productive use.

Other setbacks that could happen refer to corruption by gatekeepers or administrative agencies such as local banks and postal officials who may abuse their power and divert payments. Government agencies could also fail to channel resource revenues to the account. The author recommends setting up local ombudsman services to curtail such problems so that investigations may be conducted if anyone fails to receive their entitled amount. One of the other more long-term ways to deal with corruption is to ensure that public outrage remains sufficiently strong. In his concluding remarks, Sandbu stresses that the national wealth account is the answer to solving the resource curse and that with so much international pressure and attention to the problem, it is time to give this alternative proposal a try. However, it is not clear how all citizens or residents of all generations would receive the same real benefit over time, even after the resource is fully depleted.

Some countries set up separate funds to address the two main issues that arise from the depletion of exhaustible natural resources: uncertainty of revenues (and thus, difficulty of in budgetary planning), and intergeneration equity. To address these issues some funds were established with the objective to reduce the volatility of natural resource revenue income (the "Stabilization Fund") while others have, as an objective, the saving of current natural resource revenue for the benefit of future generations (the "Savings Fund").

While production and export diversification away from oil are necessary, so is government revenue stabilization and diversification. In most of the major oil-exporting countries that are OIC members, oil (gas) provides the lion's share of revenues and it has fluctuated considerably from year to year. Relatively stable fiscal revenues are essential for macroeconomic management and in turn for sustained economic growth. Countries that achieve sustained long-term growth on average experienced less volatility in growth:[6]

> Developing countries experience a year of negative per capita growth roughly once every three years whereas in East Asia, the average is one-half that rate and, in OECD countries, one-third that rate.

To stabilize oil revenues, countries can, and have, adopted some form of an oil stabilization fund. A portion of revenues is placed into the fund in a year of above a calculated average (moving or otherwise) expected oil revenues/prices and can be "theoretically" drawn down when revenues/prices fall below the average. Alternatively, an oil-exporting country could hedge its exposure to oil price volatility through the futures market. While such funds and hedging may be used to stabilize available oil revenues from year to year, they do not diversify the basic source of government revenues.[7] Government revenue

6. Zagha, Nankani, and Gill (2006).
7. The exception are countries that are so rich that they can invest a large portion of current oil revenues in diversified assets (abroad) to give the government all the revenue it needs in the future without having to resort to taxation.

diversification ultimately requires a healthy and growing economy with an effective income tax system.

A brief description of a few selected examples follows:

1. Norway—The State Petroleum Fund (SPF) was established in 1990 to preserve national wealth from oil production and finance the overall budget. All net income from oil flows into the SPF, from which an annual transfer is made to the treasury to meet the non-oil deficit in the budget. The fund is effectively a government account, set to provide a long-term framework for the annual process of settling the non-oil budget deficit. The fund does not directly deal with the issues of the volatile oil prices, as they are dealt with in the context of the standard budgetary process. SPF will in part also benefit future generation of Norwegians.
2. Alaska—The Alaska Permanent Fund (APF) is a State institution. APF deposits 25% of oil and gas royalty revenues into a savings fund regardless of oil market developments. The principal of the fund is invested in perpetuity, thereby transforming exhaustible natural resource into a renewable source for future generations. Annually, the principal of the fund is first replenished for inflation. Then annual distributions are made to every resident of Alaska (one year qualification), no matter what age (for minors the payment is made to the parents or guardian). As of November 18, 2016, the market value of the well-diversified fund stood at $54.5 billion.[8]
3. Chile—The Chile Copper Stabilization Fund (CSF) was created in 1985, following a sustained increase in the international copper price. The accumulation and withdrawal rules were based on a reference copper price determined annually by the authorities. The CSF was dissolved in 2002 due to significant withdrawals that began in 1998 with the sharp downturn in copper prices. However, the fund proved to be a great buffer to the expenditure pressures during the upswings of late 1980s and mid-1990s.
4. Kuwait—The Kuwait Reserve Fund for Future Generations was established in 1976 with the objective of saving in order to provide for the future generations with the depletion of resources. The fund receives 10% of total government revenue, accumulating along with

8. Retrieved November 21, 2016 from http://www.apfc.org/home/Content/home/index.cfm

the investment returns on its assets. While the fund has no precise or established rules for withdrawals, the financing of the reconstruction of Kuwait after the First Gulf War, approved by the national assembly, was a major beneficiary.

5. Oman—The Oman State General Reserve Fund was founded in 1980 with the objective of saving for future generations, receiving 15% of oil revenues. However, since its inception, the fund has become a "Contingent Fund" with the government using its discretion to withdraw from the fund to provide for budgetary deficits. Such arrangements allow for accumulation of revenues only when oil revenues exceed the budget deficit.

Askari, Abbas, Jabbour, and Kwon (2006) suggest establishing an oil fund for future generations in order to prevent waste of the nation's wealth and affording every resident of all generations an equal benefit from the transformation of oil (and gas) wealth. The authors coin their fund the Oil Fund for All Generations (OFAG) in each of the oil-exporting countries of the Persian Gulf. All of the country's revenues will be channeled into the OFAG to be managed transparently by an independent entity made accountable to the citizens of the country. The citizens will in turn, receive an annual check of constant purchasing power and borrowings from the OFAG by the government would be closely monitored. The fund would be in accordance with Islamic teachings that all humanity (in this case citizens of the country as other countries are unlikely to share their resource benefits), reduce corruption and waste and minimize the size of the public sector. The authors estimate that if all of the oil exporting countries in the Persian Gulf were to establish such a fund each country's per capita constant real payout would be significant and improve the lives of citizens in all these countries. It is interesting to note that Saudi Arabia is in fact contemplating a significant sovereign wealth fund that would directly receive a portion of the oil revenues. But unfortunately, it is a fund that does not connect to individual Saudi citizens through an annual payout of constant purchasing power.

Broadly speaking the absolute annual per capita payout figures from the proposed funds above and as a percentage of GDP per capita are very impressive and could make a tremendous difference in the lives of the average citizen. But there are significant variations between the six major oil exporters of the Persian Gulf. The six countries fall more or less into three categories: (i) Iran, (ii) Iraq and Saudi Arabia, and (iii) Kuwait, Qatar and the UAE.

Iran has the lowest payout figures. In the case of Iran, the payout ranges from 10 to nearly 40% of GDP per capita. Iran's population is large (more than the other five countries combined while Iran's ex-patriot population is relatively small) and its oil output is less than twice Kuwait's or the UAE's. Iran's payout potential could improve dramatically if it begins to exploit more aggressively its natural gas resources, as has Qatar. But still even our currently projected payout would improve the lives of the average Iranian compared to the status quo of government spending (financed from Iran's oil and gas depletion) accompanied by waste and corruption. Iraq is a unique case. It is a country that has gone back to the 1950s and is essentially starting all over again. It appears poorer than Iran because it has little in the way of modern infrastructure but it is richer than Iran in terms of per capita endowment of oil. In our opinion, the payout figures in our table for Iraq are an underestimate if Iraq can move towards peace and improved governance. Iraq could approach the category of a Saudi Arabia in terms of oil revenues per capita; and Iraq can learn from the economic policy mistakes of Iran and Saudi Arabia.

For Saudi Arabia, the payout potential is significantly higher than that of Iran or even Iraq (given Iraq's current status). The annual payout in Saudi Arabia would dramatically improve the lives of the average citizen, as the Saudi government has not delivered on economic growth and widespread prosperity. At the same time, the OFAG would provide a better guarantee of economic benefits and prosperity to future generations.

Kuwait, Qatar and the UAE are in another league. Simply said, they are rich beyond belief. Their annual payouts, even under the most conservative assumptions, are just staggering. Several items are worth pointing out in the case of these three countries. In the case of the UAE the numbers are simply staggering. Most of these funds do not belong to the entire UAE, but belong only to the Emirate of Abu Dhabi. They are all effectively millionaires based on their existing foreign investments alone! Moreover for the UAE, if the emirate of Abu Dhabi were taken by itself, the payout figure would be three times that of the UAE as a whole. If the income from existing investments were added to current (and projected) oil revenues, then the annual payouts for the UAE would increase significantly. Qatar, with its rapidly growing gas revenues and citizenry of about 200,000, is likely be in the same fortunate position as Abu Dhabi. Kuwait still has significant foreign investments, even though it spent a large portion for its liberation and reconstruction. In the case of Kuwait, Qatar and the UAE the issue is not so much that citizens of these countries are likely to starve anytime soon. Instead the issue is that the wealth of nationals should be preserved for them and for future generations in an *optimal, equitable* and *transparent* manner. This depleting wealth should not be seen as the birthright of rulers to use in order to buy loyalty, to waste on grandiose projects and military hardware, and to support shortsighted economic policies.

Looking at the spectrum of countries, in Kuwait, Qatar and the UAE, such a fund would make each and every family rich beyond belief while limiting the benefits of the ruling classes. In Saudi Arabia, and to a lesser degree Iraq, payments from the fund can make a significant contribution to the income of average families while again reducing the benefits of the rulers in Saudi Arabia and the emerging elite in Iraq. In the case of Iran the payment, though smaller, will make a difference for the average family and reduce corruption and the take of the elite. All of this will be most effective if it is supported by an effective income tax system and a period of transition as government oil revenues are reduced to zero. Careful consideration needs to be given in each country, on a country-by-country basis, to design a system that affords appropriate incentives to individuals to live productive lives and to contribute to national economic and social prosperity.

We should emphasize that the annual transfers from such a fund to all citizens would be for all generations and for all time. In this way, the capital in the ground in the form of oil and gas would be transfomed into diversified financial assets that would continue to be the patrimony of all citizens. In the process, rent-seeking activities would be diminished and support more effective insitutions. Furthermore, we would venture to add that such a fund would reduce the incentives for those in power and their supporters to continue their oppressive rule.

Another suggestion raised by Acemoglu et al. (2005) to enable good economic institutional outcome is to establish a higher political institution that would place checks and balances on those holding political power. In their view, without this political institution, power holders are more likely to misuse the power and opt for policies that would prove to be beneficial for them. As a result of this, property rights of the major population will not be protected and societies suffer. The authors also stress that institutional weaknesses can be rectified when there are limited rents that power holders can extract from society.

The suggestions provided so far are new recommendations to manage resource funds. There is, however, an initiative that has been set up to mitigate the negative effects of resource abundance and address institutional weaknesses and referred to as the Extractive Industries Transparency Initiative (EITI). The EITI is an international policy intervention that aims to promote the transparency of resource revenues and government accountability. It was first launched in 2002 by U.K. Prime Minister Tony Blair with 33 country members designated as EITI compliant (Corrigan, 2014). Corrigan (2014) explains that the EITI is an organization that seeks to increase transparency through the disclosure of company payments and government revenues, as well as to curb corruption. Participation by a government in the EITI is seen as willingness to reform and validates the EITI's international standards. According to Corrigan (2014), countries that do not accede to these standards would face decreasing international legitimacy as they are at risk of losing their reputation, which may also impact their credit rating, foreign investment or international financial aid (if any). The EITI requires countries to publish reports critical to government and industry practices; thus in theory it should reduce the severity of negative governance practices, strengthen connected institutions and mitigate negative effects that would typically affect resource-rich countries (Corrigan, 2014).

The second part of the study conducted by Corrigan (2014) reviews the success of the EITI in mitigating the resource curse by examining two common outcomes negatively affected by resource abundance. These include economic development and the quality of governance. The author notes that although EITI has shown early signs of positive impact in financial reporting, the organization also faces several limitations. First, it does not show enough visibility and governments have yet to play a full and free part in the EITI. Secondly, the validation process conducted by EITI itself poses conflict of interest issues and should instead be conducted by an independent body. Finally, the EITI has yet to include disaggregate reporting of each country's revenues with company-to-company earnings breakdown for further public accountability.

Corrigan also cites past research that scrutinizes the minimum standards of EITI as it is deemed inefficient due to the findings that EITI candidates are still reporting inadequate and inaccurate revenues. Another concern refers to the membership structure of the EITI. As it is voluntary in nature, enforceability will be difficult. Nonetheless, the author's review on the effectiveness of EITI provided the following results:

1. EITI membership brought a positive effect in cases where resource abundance is substantial and significant for a country.

2. A significant increase in GDP per capita since membership is noted which could indicate a response to more development-oriented policies.
3. Citizen's trust on institutions has not improved and means of informing citizens has not been adequate. Citizens are seen as unaffected by any improvements in transparency levels.
4. Any improvement in corruption is seen as sporadic, indicating that the EITI's effect on corruption is weak and that it does not act as a deterrent to curb this behavior.
5. EITI's membership is seen as more to improve transparency rather than accountability as the effects on political stability and lack of violence indicators were insignificant. In respect of this, the EITI is seen as unsuccessful in eliminating counterproductive behaviors such as greed associated with resource abundance.

The study also found that countries that do not have sufficient resources do not experience benefits from joining the EITI. The author stresses that the results of the study combine both positive and negative effects. A repeated study is proposed after a number of years to reassess the impact on membership as the impact of governance may take a longer duration to assess. Thus at this stage, the effectiveness of the EITI can be seen in improving transparency, but it remains questionable in reducing corruption and helping countries achieve goals of stable democratic systems (Corrigan, 2014). Essentially, resource-rich countries struggle more than others to keep checks and balances in place and that solution should be tailored according to the social, economic and political milieus of the respective countries (Corrigan, 2014).

The appropriate way to view the nature, manifestation and exploitation of depletable resources was illuminated by Robert Solow over 50 years ago and was aptly summarized as:

> The finite pool of resources (I have excluded full recycling) should be used up optimally according to the general rules that govern the optimal use of reproducible assets. In particular, earlier generations are entitled to draw down the pool (optimally, of course!) so long as they add (optimally, of course!) to the stock of reproducible capital.[9]

9. Solow (1974).

Islam from its birth, much earlier and centuries ago, preached that natural resources were The Almighty's gift to humanity. These resources were to benefit all humanity. Natural resources are the heritage of all individuals of every generation, and current generations must afford equal benefit to those who cannot access natural resources that are Allah's Gift and also preserve the right of future generations—for every individual in each generation to reap the same benefit.

Robert Solow reached a similar conclusion to what Islam preaches—depletable resources are an important part of the capital that humanity has been given. It must not be consumed. It can, however, be transformed into other forms of capital. This transformation must be done efficiently. The benefits must be shared equally by all humans of all generations. Solow's only departure from the Islamic prescription is on the final point, the accrual of equal benefits to all humans of all generations.

Conclusion

In summing up, one can draw a number of important conclusions from the discussions that motivated our study. There is ample empirical research to indicate that institutions are at the foundation of sustained economic growth, especially in natural resource-abundant countries. As a result, we will expand on the strand of contributions that indicate the importance of the quality and effectiveness of institutions as explanations for the presence (or absence) and strength of the resource curse.

It is worth noting that none of the earlier research on the resource curse has focused on country groups of resource-rich countries such as the OIC, where the commonality of belief, structural form and policy goals could provide a more universal understanding on economic performance and behavior. As a result, none of the existing studies has looked at Islamic principles and teachings as a way to manage depletable natural resources through the principles of justice and equal distribution of benefits, which would be the Islamic prescription.

References

Acemoglu, D., Johnson, S., & Robinson, J. a. (2005). Chapter 6 institutions as a fundamental cause of long-run growth. *Handbook of Economic Growth*, *1*(SUPPL. PART A), 385–472. https://doi.org/10.1016/S1574-0684(05)01006-3

Al Rawashdeh, R., & Maxwell, P. (2013). Jordan, minerals extraction and the resource curse. *Resources Policy*, *38*(2), 103–112. https://doi.org/10.1016/j.resourpol.2013.01.005

Algieri, B. (2011). The Dutch Disease: Evidences from Russia. *Economic Change and Restructuring*, *44*(3), 243–277. https://doi.org/10.1007/s10644-011-9101-4

Apergis, N., & Payne, J. E. (2014). The oil curse, institutional quality, and growth in MENA countries: Evidence from time-varying cointegration. *Energy Economics*, *46*, 1–9. https://doi.org/10.1016/j.eneco.2014.08.026

Askari, H. (1990). *Saudi Arabia's economy: Oil and the search for economic development*. Stamford, CT: JAI Press.

Askari, H. (2006). *The Middle East oil exporters: What happened to economic development?* Cheltenham: Edward Elgar Publishing.

Askari, H. (2012). *Conflicts and wars: Their fallout and prevention*. New York, NY: Palgrave Macmillan.

Askari, H., Abbas, F., Jabbour, G., & Kwon, D. (2006). An economic manifesto for the oil exporting countries of the Persian Gulf. *Banca Nazionale Del Lavoro Quarterly Review*, *59*(239), 363–388. Retrieved from http://ovidsp.ovid.com/ovidweb.cgi?T=JS&CSC=Y&NEWS=N&PAGE=fulltext&D=econ&AN=0901175

Auty, R. M. (2007). Natural resources, capital accumulation and the resource curse. *Ecological Economics*, *61*, 627–634. https://doi.org/10.1016/j.ecolecon.2006.09.006

Ayelazuno, J. (2014). Oil wealth and the well-being of the subaltern classes in Sub-Saharan Africa: A critical analysis of the resource curse in Ghana. *Resources Policy*, *40*, 66–73. https://doi.org/10.1016/j.resourpol.2013.06.009

Basedau, M., & Lay, J. (2009). Resource curse or rentier peace? The ambiguous effects of oil wealth and oil dependence on violent conflict. *Journal of Peace Research*, *46*(6), 757–776. https://doi.org/10.1177/0022343309340500

Beine, M., Bos, C. S., & Coulombe, S. (2012). Does the Canadian economy suffer from Dutch disease? *Resource and Energy Economics*, *34*(4), 468–492. https://doi.org/10.1016/j.reseneeco.2012.05.002

Bjorvatn, K., Farzanegan, M. R., & Schneider, F. (2012). Resource curse and power balance: Evidence from oil-rich countries. *World Development*, *40*(7), 1308–1316. https://doi.org/10.1016/j.worlddev.2012.03.003

Boschini, A., Pettersson, J., & Roine, J. (2013). The resource curse and its potential reversal. *World Development*, *43*, 19–41. https://doi.org/10.1016/j.worlddev.2012.10.007

Bulte, E. H., Damania, R., & Deacon, R. T. (2005). Resource intensity, institutions, and development. *World Development*, *33*(7), 1029–1044. https://doi.org/10.1016/j.worlddev.2005.04.004

Caselli, F. (2006). Power struggles and the natural resource curse. *LSE Working Paper*, (December 2005), 1–2. Retrieved from http://eprints.lse.ac.uk/4926

Cavalcanti, T. V. D. V., Mohaddes, K., & Raissi, M. (2011). Growth, development and natural resources: New evidence using a heterogeneous panel analysis. *The Quarterly Review of Economics and Finance*, *51*(4), 305–318. https://doi.org/10.1016/j.qref.2011.07.007

Cockx, L., & Francken, N. (2014). Extending the concept of the resource curse: Natural resources and public spending on health. *Ecological Economics*, *108*, 136–149. https://doi.org/10.1016/j.ecolecon.2014.10.013

Corrigan, C. C. (2014). Breaking the resource curse: Transparency in the natural resource sector and the extractive industries transparency initiative. *Resources Policy*, *40*, 17–30. https://doi.org/10.1016/j.resourpol.2013.10.003

Costa, H. K. D. M., & Santos, E. M. Dos. (2013). Institutional analysis and the "resource curse" in developing countries. *Energy Policy*, 63, 788–795. https://doi.org/10.1016/j.enpol.2013.08.060

David, G. A., & Tilton, J. E. (2005). The resource curse. *Natural Resources Forum*, *29*, 233–242. https://doi.org/10.1080/05679320500129037

Dülger, F., Lopcu, K., Burgaç, A., & Balli, E. (2013). Is Russia suffering from Dutch Disease? Cointegration with structural break. *Resources Policy*, *38*(4), 605–612. https://doi.org/10.1016/j.resourpol.2013.09.006

El Anshasy, A. a., & Katsaiti, M.-S. (2013). Natural resources and fiscal performance: Does good governance matter? *Journal of Macroeconomics*, *37*, 285–298. https://doi.org/10.1016/j.jmacro.2013.05.006

Everhart, S. S. (2010). The resource curse and private investment: A theoretical model of the impact of corruption. *Education, Business and Society: Contemporary Middle Eastern Issues*, 3(2), 117–135. https://doi.org/10.1108/17537981011047952

Glaeser, E. L., La Porta, R., Lopez-De-Silanes, F., & Shleifer, A. (2004). Do institutions cause growth? *Journal of Economic Growth*, 9, 271–303.

Hodler, R. (2006). The curse of natural resources in fractionalized countries. *European Economic Review*, 50(6), 1367–1386. https://doi.org/10.1016/j.euroecorev.2005.05.004

Humphreys, M. (2005). Natural resources, conflict, and conflict resolution: Uncovering the mechanisms. *Journal of Conflict Resolution*, 49(4), 508–537. https://doi.org/10.1177/0022002705277545

James, A., & Aadland, D. (2011). The curse of natural resources: An empirical investigation of U.S. counties. *Resource and Energy Economics*, *33*(2), 440–453. https://doi.org/10.1016/j.reseneeco.2010.05.006

Kolstad, I., & Søreide, T. (2009). Corruption in natural resource management: Implications for policy makers. *Resources Policy*, *34*(4), 214–226. https://doi.org/10.1016/j.resourpol.2009.05.001

Kolstad, I., & Wiig, A. (2012). Testing the pearl hypothesis: Natural resources and trust. *Resources Policy*, *37*(3), 358–367. https://doi.org/10.1016/j.resourpol.2012.04.003

Kutan, A. M., & Wyzan, M. L. (2005). Explaining the real exchange rate in Kazakhstan, 1996–2003: Is Kazakhstan vulnerable to the Dutch disease? *Economic Systems*, *29*, 242–255. https://doi.org/10.1016/j.ecosys.2005.03.009

Larsen, E. R. (2005). Are rich countries immune to the resource curse? Evidence from Norway's management of its oil riches. *Resources Policy*, *30*(2), 75–86. https://doi.org/10.1016/j.resourpol.2004.12.001

Latif, S., Farooq, A., Loganathan, N., & Shahbaz, M. (2014). Empirical evidence on the resource curse hypothesis in oil abundant economy. *Economic Modelling*, *42*, 421–429. https://doi.org/10.1016/j.econmod.2014.07.020

Le Billon, P. (2001). The political ecology of war: Natural resources and armed conflicts. *Political Geography*, *20*(5), 561–584. https://doi.org/10.1016/S0962-6298(01)00015-4

Mehlum, H., Moene, K., & Torvik, R. (2006). Institutions and the resource curse. *Economic Journal*, *116*(508), 1–20. https://doi.org/10.1111/j.1468-0297.2006.01045.x

Mideksa, T. K. (2013). The economic impact of natural resources. *Journal of Environmental Economics and Management*, 65(2), 277–289. https://doi.org/10.1016/j.jeem.2012.07.005

Paine, T. (1995). *Thomas paine collected writings*. New York, NY: Library of America.

Pegg, S. (2010). Is there a Dutch disease in Botswana? *Resources Policy*, 35(1), 14–19. https://doi.org/10.1016/j.resourpol.2009.07.003

Robinson, J. a., Torvik, R., & Verdier, T. (2014). Political foundations of the resource curse: A simplification and a comment. *Journal of Development Economics*, *106*, 194–198. https://doi.org/10.1016/j.jdeveco.2013.09.004

Ross, M. L. (2004). What do we know about natural resources and civil war? *Journal of Peace Research*, *41*(3), 337–356. https://doi.org/10.1177/0022343304043773

Sachs, J. D., & Warner, A. M. (1995). Natural Resource Abundance and Economic Growth. *NBER Working Paper Series*, 3, 54. https://doi.org/10.3386/w5398

Sachs, Jeffrey and Andrew Warner, (1997), "Natural Resource Abundance and Economic Growth", mimeo, Center for International Development, Harvard University. (1997).

Sachs, J. D., & Warner, A. M. (2001). The curse of natural resources. *European Economic Review*, 45(4–6), 827–838. https://doi.org/10.1016/S0014-2921(01)00125-8

Sandbu, M. E. (2006). Natural wealth accounts: A proposal for alleviating the natural resource curse. *World Development*, 34(7), 1153–1170. https://doi.org/10.1016/j.worlddev.2005.11.014

Solow, R. M. (1974). Intergenerational equity and exhaustible resources. *Review of Economic Studies*, p.41.

Stevens, P., & Dietsche, E. (2008). Resource curse: An analysis of causes, experiences and possible ways forward. *Energy Policy*, 36, 56–65. https://doi.org/10.1016/j.enpol.2007.10.003

Sala-i-Martin, X., & Subramaniam, A. (2003). Addressing the Natural Resource Curse: An Illustration from Nigeria. *IMF Working Paper, WP/03/139*.

UCDP. (2010). Conflict Encyclopedia. Uppsala University. https://ucdp.uu.se/database

Van der Ploeg, F. (2010). Why do many resource-rich countries have negative genuine saving? *Resource and Energy Economics*, 32(1), 28–44. https://doi.org/10.1016/j.reseneeco.2009.07.002

Watts, M. (2010). Resource curse? Governmentality, oil and power in the Niger Delta, Nigeria. *Geopolitics*, 9(1), 50–80. https://doi.org/10.1080/14650040412331307832

Williams, A. (2011). Shining a light on the resource curse: An empirical analysis of the relationship between natural resources, transparency, and economic growth. *World Development*, 39(4), 490–505. https://doi.org/10.1016/j.worlddev.2010.08.015

Yuxiang, K., & Chen, Z. (2011). Resource abundance and financial development: Evidence from China. *Resources Policy*, 36(1), 72–79. https://doi.org/10.1016/j.resourpol.2010.05.002

Zagha, R., Nankani, G., & Gill, I. (2006 March). *Rethinking growth. Finance and development* (p. 8). Washington, DC: International Monetary Fund.

· 3 ·

RESOURCE CURSE IN OIC COUNTRIES

The Role of Institutions

Introduction

In order to address the presence of resource curse for the oil-producing OIC countries, we adopt a growth regression framework typically employed in the resource-curse literature. Yuxiang and Chen (2011) referred to this model as the "resource-growth model." More specifically, the models follow the resource growth strand of studies that have looked into institutions as a determinant of economic growth in countries with abundant depletable natural resources (see Bjorvatn, Farzanegan, & Schneider, 2012; Costa & Santos, 2013; El Anshasy & Katsaiti, 2013; Latif, Farooq, Loganathan, & Shahbaz, 2014; Mehlum, Moene, & Torvik, 2006; Oskenbayev, Yilmaz, & Abdulla, 2013; Tsani, 2015) The base model is as follows:

$$\text{Ln } Y_{it} = \mu_i + \beta_1 \ln \text{Oil} R_{it} + \beta_2 \, INST_{it} + \beta_3 \ln FD_{it} + \theta X_{it} + e_{it}$$

Model 1

Where *Ln Y* is the log of real GDP per capita, *ln OilR* is the log of oil rent, *INST* is a measure of institutional factors, *ln FD* is the log of the measure of financial development, and *X* stands for the control variables and refers to the error term. *i* and *t* respectively represent country and time. The key variable of interest in the model is oil rent. Logically oil revenues would affect GDP positively. However, it is well noted that some countries with abundance of resources are observed to progress slower due to the resource curse. Accordingly, if β_1 is greater than zero, it shows that oil rent contributes positively to GDP. In this regard, the resource curse is not present. A negative β_1 is indicative of an inverse association between oil revenue and GDP and confirms the presence of the resource curse. The negative relationship is in line with theoretical predictions by past researchers, suggesting that resource abundance can slow down economic progress (see Auty, 2007; Ayelazuno, 2014; James and Aadland, 2011; Larsen, 2005 and Latif et al., 2014).

Earlier work by Sachs and Warner and others revealed that GDP is not independent of institutional quality (Bulte, Damania, & Deacon, 2005). Guided by empirical results from Sachs and Warner, the variable for institutions is built into the model as a second determinant of GDP. Stronger institutions are indications that governance is sound and transparent and that corruption or rent-seeking activities are minimal. Studies by Knack and Keefer (2012), Brunnschweiler (2008), Glaeser, La Porta, Lopez-De-Silanes, and Shleifer (2004) are examples in the literature that postulate institutions have a positive relationship with GDP. More specifically, stronger institutions often trigger more sound expenditure and developmental policies (see Brunnschweiler & Bulte, 2008; El Anshasy & Katsaiti, 2013; Knack & Keefer, 2012).

On the other hand, weaknesses or fractionalized institutions affect economic growth negatively. According to El Anshasy and Katsaiti (2013), one of the explanations for this adverse relationship can be seen through misappropriation of funds by powerful groups that compete to appropriate a greater share of national wealth for the benefit of their constituencies. Thus β_2 is expected to have a positive sign.

Financial development has been demonstrated in the literature to be an important mechanism for long-run growth (see Baltagi, Demetriades, & Law, 2009; Levine, 2000; Yuxiang & Chen, 2011). Baltagi et al. (2009) suggests that economic growth is encouraged when financial development creates opportunities for new firms to become established. In a similar study, Levine (2000) suggests that the linkage between financial development and growth happens when stock markets liquidity is enhanced by the improvement of domestic financial markets, thereby leading to a boost in productivity growth. Based on these explanations, we expect financial development to have a positive relationship with economic growth ($\beta_3 > 0$).

The control variables in the model are other variables thought to be important determinants of GDP. Following Bjorvatn et al. (2012) the control variables employed include investments, government final consumption expenditure ("government expenditure") and trade. Investment is considered as an essential feature for growth of an economy as posited by Solow's neoclassical growth model (Prescott, 1956). As explained by Prescott (1956), Solow's application of the model provides evidence that co-movements of variables, which included investments, capital and labor input was able to provide explanations on understanding economic growth and its patterns. Further contributions by other researchers, who also contribute to the search for a set of robust variables to model growth, have identified investment as part of its "core explanatory variables" (Li & Liu, 2005). Based on these studies, investments have been found to have a positive effect on economic growth and as such, we expect the same results.

In relation to government expenditures, El Anshasy and Katsaiti (2013) and Ahmed and Miller (1999) explain that a number of studies have found the government expenditure model to have a negative or insignificant effect on growth. Nevertheless, government expenditure has been seen to affect growth through investments according to the "crowding in and out" effect. According to Ahmed and Miller (1999), the "crowding out" view argues that government expenditure crowds out private investment and that higher government investment expenditure increases demands for goods and services, raises interest rates making capital more expensive and as such reducing private investment. The non-traditional "crowding in" view is that government expenditure can induce investment. In consideration of these theories and predictions, we expect that government expenditure could generate either a positive or a negative sign. Finally, we employ trade openness as a third control variable. Edwards (1992) explains that countries that have liberalized external sectors and reduced their impediments to international trade will outperform countries that do not do so. According to Edwards, a number of studies in the economics literature see trade openness as encouraging economic growth. Similar results were also reported by Bjorvatn et al. (2012) and Yuxiang and Chen (2011). In light of this, we expect trade to have a positive relationship with GDP.

The preceding discussions have thus far provided the setting to establish the presence of the resource curse by capturing the independent influences of the variables mentioned. We then extend the resource curse analysis by investigating whether the effect between oil rent and economic growth is conditional upon the level and quality of institutions. In other words, we assess whether improving the level of institutions will have any bearing on how oil rent affects GDP. Thus, we shall include the interaction term *lnOil-Rent x Inst* into Model 1 and term the model as Model 2 as follows:

$$\text{Ln } Y_{it} = \mu_i + \beta_1 \ln \text{Oil} R_{it} + \beta_2\, INST_{it} + \beta_3 \ln FD_{it} + \lambda_1 (\ln \text{Oil} R_{it} \times INST_{it}) + \theta X_{it} + e_{it}$$

Model 2

As can be seen in the model, all variables remain as mentioned in Model 1 above. Recall the presence of resource curse can be seen if β_1 is negative. The interaction term in this model is captured by the parameter λ_1. A negative β_1 coupled with a positive sign for λ_1 would indicate that the presence of the resource curse is conditional on the level of institutional quality as:

$$\ln Oil\, R_{it} \beta_1 + \ln \lambda_1\, INST_i$$

Equation i

In the case of Model 2, this means that increasing oil rents with weak institutions causes detrimental effects on income. In other words, the negative sign for β_1 with a positive sign for λ_1 indicate the resource curse is stronger if the quality of institutions is poor. We can also derive the institutional threshold, λ^* from Equation i as:

$$\lambda^* = \lambda 1$$

Equation ii

The threshold tells us the level at which the resource curse is neutralized for the countries under study. We predict that if the value of institutions is lower than the threshold (INST < λ^*), then the resource curse is strong and resource abundance is harmful to growth. If it is above the threshold (INST > λ^*), the effect of the resource curse does not apply, and resource abundance becomes a contributor to growth. A negative β_1 and a non-significant or negative λ_1 on the other hand indicate that improving on institutional quality will not decrease the resource curse.

Institutions and Oil Rent Relations

In the preceding discussions, we would establish in Models 1 and 2 whether the resource curse exists and whether the resource curse is conditional upon the quality of institutions. We then examine further the role of institutions in relation to the resource growth association and evaluate in this section whether oil rent affects GDP indirectly through its impact on institutions.

In Models 1 and 2, we have maintained GDP as the dependent variable. Our prediction is that the presence of the resource curse could manifest itself through other variables that contribute to economic growth. We predict that the institution variable is an important determinant of economic growth, and we examine the relationship with oil rent to determine the existence of the resource curse. For this purpose, Model 3 is constructed with institutions as the dependent variable:

$$INST_{it} = \mu_i + \beta_1 \ln OilR_{it} + \beta_2 \ln Y_{it} + \beta_3 \ln INV_{it} + \beta_4 \ln TRADE_{it} + \theta X_{it} + e_{it}$$

Model 3

Where *INST* is a measure of institutional factors, *ln OilR* is the log of oil rent (as a share of GDP), *ln Y* is the log of real GDP per capita, *ln INV* is the log of investments, *ln TRADE* is the log of trade openness and *X* stands for the control variables. These determinants were adopted from the work of Kolstad and Wiig (2012). The test is focused first on establishing the direct effect between oil rent and institutions, i.e., an effect that is positive indicates that resource abundance contributes to improvement in the quality of institutions. A negative relationship between oil rent and institutions on the other hand indicates that the increase in oil rent causes erosion in the quality of institutions. The latter indicates the presence of the resource curse. Based on the literature, the negative relationship between oil rent and institutions has been attributed to rent-seeking opportunities that arise as a result of resource abundance (Bulte et al., 2005).

However, the expected relationship between oil rent and institutions remains ambiguous due to the mixed results from past studies. For example, Brunnschweiler (2008) found no evidence that resource abundance affects institutions negatively, thus contradicting the resource curse theory. Other studies such as Oskenbayev et al. (2013), Sachs and Warner (1997), Bulte et al. (2005) found that countries with abundant resources end up with worse institutions. In consideration of these contrasting results the sign for β_1 could be positive or negative. If β_1 is negative based on the indirect linkage of the resource institutions relations with economic growth we can infer that due to the resource curse, resource abundance causes erosion in quality of institutions, which, in turn, causes detrimental effects to economic growth. The results would thus suggest the ability of improving economic growth and reducing the resource curse through improving the quality of institutions.

The relationship between investment and institutions in Model 3 is expected to be positive. According to Oskenbayev et al. (2013) this implied effect stems from the fact that higher investments creates pressure for governments to provide better institutions. The same relationship is expected for trade as according to Oskenbayev et al. (2013) trade escalates better regulation and as a result demoralizes rent-seeking behaviors and corruption. The control variables were selected based on a similar resource-curse study that had assessed the relationship between institutions and the resource-curse (see Kolstad & Wiig, 2012). Kolstad and Wiig (2012) selected the control variables unemployment rate, school enrolment at the tertiary level and population size as measures of equal opportunity. We adopt these same variables and expect the same positive results found by their research.

Financial Development and Oil Rent Relations

Recent theories suggest that financial development could have a bearing on how oil rent affects GDP and that improving the quality and deepening of financial development will have a bearing on ameliorating the fallout of the resource curse.

We thus extend our analysis and assess how resource abundance affects financial development. Our main contention is that although resource abundance affects financial development, a comparison of the impact of financial development and the quality of institutions, the impact of the quality of institutions is significantly greater on economic growth and thus in alleviating the fallout of the resource curse. Model 4 is adopted below and financial development is treated as the dependent variable:

$$\ln FD_{it} = u_i + \beta_1 \ln Oil\, R_{it} + \beta_2 \ln INV_{it} + \theta X_{it} + e_{it}$$

Model 4

The variable of interest remains the oil rent variable included as a first determinant in the model. According to Yuxiang and Chen (2011), although there have been a number of studies in the past linking financial development and economic growth, research to evaluate resource abundance and financial development is still nascent. Nonetheless, the authors found a negative relationship between them, thereby confirming the resource curse hypothesis. In their opinion, resources contribute to financial development positively; however, due to the exploitation of resources the potential for financial development becomes affected. This same reasoning is adopted to hypothesize the expected relationship between oil rent and financial development in Model 4. A negative relationship signals the presence of the resource curse and confirms that resources do not contribute positively to financial development.

In selecting the control variables, Yuxiang and Chen (2011) state the importance of trade openness in financial development in previous studies (see also Baltagi et al., 2009). The authors also highlight the use of Foreign Direct Investment (FDI) as a percentage of GDP to account for financial openness. In theory, the authors state that lack of financial openness would have an effect on the strength of financial development. For the third control variable, Yuxiang and Chen (2011) introduced the interaction term between trade and FDI. Here the authors state in theory that simultaneous opening of both trade and capital accounts will have an impact on financial development. A three-year lagged value of these variables is used in the model in accordance with Yuxiang and Chen (2011) as the authors indicate the need to avoid any bias due to the reverse causality from financial development to trade or FDI.

Similarly, Baltagi et al. (2009) also treated these variables the same way when analyzing the relationship between resource abundance and financial development. A significant and positive relationship for the variables investment and trade would indicate that they affect financial development in a positive manner.

Data

The dataset used in our study covers 10 oil-producing countries.[1] The other OIC countries that were listed as oil-producing and oil-exporting countries were omitted because data was not available for all the core variables to be tested across the time period chosen.[2] The beginning year for the analysis is 1984 because this was the first year data was available for all countries under study.

1. The 10 countries in the sample are: Algeria, Bahrain, Cameroon, Cote D'Ivoire, Egypt, Indonesia, Malaysia, Nigeria, Oman and Saudi Arabia.
2. Oil-exporting country listing according to the United States energy information administration www.eia.gov

Following Bulte et al. (2005), we use the real GDP per capita data from the World Bank database for the growth variable. The variable oil rent relates to the difference between the value of crude oil production at world prices and total costs of production. According to Yuxiang and Chen (2011) three kinds of resource proxies, estimated on the basis of resource export, reserve or production data was usually used by previous researchers in their resource-curse model. The study conducted by Sachs and Warner (1995) measured resource export as a percentage of GDP and became the most commonly used measure in previous studies (Yuxiang and Chen, 2011). Stijns (2005), however, argues that the correlation between reserve data and primary exports is weak.

Similarly, El Anshasy and Katsaiti (2013) found that the use of share of resource exports to a country's total exports is dependent on the structure of the economy. They thus adopt instead the use of resource rent data. The resource rent data, in particular, oil rent data from the World Bank database, will be used across the four models for our study based on this reason.

The proxy for the institutional variable is made up of various measures of institutions and governance qualities available from the International Country Risk Guide (ICRG) data. These measures relate to the average of six indicators from the ICRG comprising of the government stability, corruption quality, law and order, democratic accountability and bureaucratic quality indexes. The ICRG data set is produced by the PRS Group, Inc., a UK based firm producing globally recognized indicators and systems. El Anshasy and Katsaiti (2013) were in support of this data and stated that it presents wide country coverage for a reasonably long series. This variable was also the underlying composite measure of "institutional quality" used in Boschini, Pettersson, and Roine (2013) in their resource-curse and growth study. We alternate between the money supply and the domestic credit to private sector data from the World Bank database as the measure for financial development. According to Latif et al. (2014), Bjorvatn et al. (2012) and Yuxiang and Chen (2011) these are common proxies to measure financial development. We included the investment, government expenditure and trade openness variables as control variables for Models 1 and 2.

Investment refers to the countries' gross domestic investment data taken from the World Bank database and government expenditure data from the same source refers to all government current expenditures. The measure for trade openness relates to the ratio of the sum of exports and imports to GDP. For Model 3, we control for determinants of the level of institutional quality by employing the unemployment, population size and tertiary education enrolment variables often found in the literature (see Kolstad & Wiig, 2012; Mehlum et al., 2006; Tsani, 2015).

In Model 4, three variables are incorporated to control the effects of other factors on financial development. First, the ratio of the sum of exports and imports to GDP is estimated as the trade openness variable. The second control variable relates to financial openness and is measured by FDI and finally according to Yuxiang and Chen (2011) the interaction term between trade and FDI is also introduced as the third control variable.

These variables were obtained from the World Bank database. Table 3.1 provides additional details on the data and its sources.

Table 3.1: Data Description.

No.	Notation	Variable	Definition[3]	Source
1	Y	GDP	Gross Domestic Product in US Dollars. GDP at purchaser's prices is the sum of gross value added by all resident producers in the economy plus any product taxes and minus any subsidies not included in the value of the products.	World Bank Database
2	Oilrent	Oil Rent	Oil rents are the difference between the value of crude oil production at world prices and total costs of production.	World Bank Database
3	INST	Institutional Quality	Institutional Quality Indicator comprising the average of 5 Indicators which are Government Stability, Corruption Quality, Law and Order, Democratic Accountability and Bureaucratic Quality; taken from International Country Risk Guide.	International Country Risk Guide Database
4	FD	Financial Development	Domestic Credit to Private Sector. It measures financial resources provided to the private sector by financial corporations, such as through loans, purchases of non-equity securities, and trade credits and other accounts receivable, that establish a claim for repayment.	World Bank Database

(*Continued*)

3. Definition provided by the source database.

Table 3.1: (*Continued*)

No.	Notation	Variable	Definition[3]	Source
4	FD	Financial Development	Money supply as a percentage of GDP. Money and quasi money comprise the sum of currency outside banks, demand deposits other than those of the central government, and the time, savings, and foreign currency deposits of resident sectors other than the central government.	World Bank Database
5	Inv	Investment	Gross capital formation (formerly gross domestic investment) consists of outlays on additions to the fixed assets of the economy plus net changes in the level of inventories.	World Bank Database
6	GCE	Government Consumption & Expenditure	General government final consumption expenditure (formerly general government consumption) includes all government current expenditures for purchases of goods and services (including compensation of employees). It also includes most expenditure on national defense and security, but excludes government military expenditures that are part of government capital formation.	World Bank Database
7	Trade	Trade	Trade is the sum of exports and imports of goods and services measured as a share of gross domestic product.	World Bank Database
8	Un	Unemployment	Unemployment refers to the share of the labor force that is without work but available for and seeking employment.	World Bank Database
9	Te	Tertiary Enrolment	Total is the total enrollment in tertiary education), regardless of age, expressed as a percentage of the total population of the five-year age group following on from secondary school leaving.	World Bank Database

Table 3.1: (*Continued*)

No.	Notation	Variable	Definition[3]	Source
10	Pop	Population Size	Total population is based on the de facto definition of population, which counts all residents regardless of legal status or citizenship—except for refugees not permanently settled in the country of asylum, which are generally considered part of the population of their country of origin. The values shown are midyear estimates.	World Bank Database
13	INST* OilRent	Oil Rent and INST	Interaction term made up of the Oil Rent and Institutional Variable	World Bank Database and ICRG
14	FDI	Foreign Direct Investment	Foreign direct investment is the net inflows of investment to acquire a lasting management interest in an enterprise operating in an economy other than that of the investor. It is the sum of equity capital, reinvestment of earnings, other long-term capital, and short-term capital as shown in the balance of payments.	World Bank Database

Estimation Methods

As a basic approach, we adopt the traditional fixed effect model to evaluate the existence of resource curse. Then, for robustness, we apply the heterogenous estimation approach, the Pooled Mean Group (PMG) and Mean Group (MG) estimators.

Basic Panel Estimation

Writing the models as given in 1 to 3 in its generic form, we have:

$$Y_{it} = \mu_i + X_{it} + e_{it}$$

Where Y refers to independent variables, X the vector of variables and *e* stands for the error term. The advantage of using the panel modelling approach is that it allows for heterogeneity across the cross-sectional unit by specifying the error term to capture individual-specific effects. The fixed effect ordinary least squares model allows the intercept to vary through a dummy variable technique, however, it assumes that the slope of the coefficient is constant. Nevertheless, although the intercept may differ across countries, each individual's intercept does not vary over time, in that it is time invariant. In other words, the fixed effect model is able to capture all effects which are specific to a particular country and which do not vary over time (Asteriou & Hall, 2007; Brooks, 2008).

The random effects model, on the other hand, expresses the lack of knowledge of the true model through the error term e_{it} instead of employing the use of dummy variables. The random effects model which is sometimes also known as the error components model proposes different intercept terms for each country and these intercepts are constant over time with the relationship between explanatory and explained variables assumed to be the same both cross-sectionally and temporarily (Brooks, 2008). Gujarati (2004) provides the following equation to explain the model further:

$$e_{it} = v_i + u_{it}$$

The composite error term e_{it} consists of two components, u_i, which is the cross section or individual-specific, error component and the combine time series and cross-section error component is denoted as u_{it}. The following are the provided assumptions of the random effects method:

$$v_i \sim N(0, \ {}_{v})$$

$$u_{it} \sim N(0, \sigma_u^{\,2})$$

$$E(v_i, u_{it}) = 0 \quad E(v_i, u_j) = 0 \quad (i \neq j)$$

$$E(u_{it}, u_{js}) = E(u_{it}, u_{jt}) = E(u_{it}, u_{js}) = 0 \qquad (i \neq j; t \neq s)$$

Thus, there is no correlation between the individual error components, and they are also not auto-correlated across cross-section and time-units. In general, the difference between the two methods is that the fixed effects model assumes that the intercept for each country differs and in random effects each country differs in its error term. Gujarati further explains that the difference between fixed effects and random effects here is that in the former each cross-sectional unit has its own (fixed) intercept value. In the latter, the intercept represents the mean value of all of the cross-sectional intercepts, where v_i represents the error components model, which refers to the random deviation of individual intercept from this mean value.

In order to decide which method between fixed and random effects to adopt, a number of considerations will have to be taken into account. Firstly, according to Gujarati (2004), this answer hinges around the assumption one makes about the likely correlation between the individual, or cross-section specific, error component model e_{it} and X regressors. The random effects method assumes that v_i and X regressors are uncorrelated because in the underlying assumption of the model, v_i is a random drawing from a much larger population. If v_i and X regressors are correlated then the fixed effects method is more appropriate. For example, in the case of modeling economic growth where we suppose our growth determinants as resources and investments, v_i stands for the unobserved effects such as technological progress. In this respect, it is very likely for v_i to be correlated with resources and investments; thus in this case the random effects method is not tenable. Moreover, Gujarati (2004) explains that studies in the literature have argued that in many applications the entire reason for using panel data is to allow for the unobserved effects to be correlated with explanatory variables.

Secondly, according to Asteriou and Hall (2007), when a panel is balanced the fixed effect method will work best. We deduce that our panel is balanced from the explanation given by Gujarati (2004) that each cross-sectional unit or observation has the same number of time series observations. Finally, these methods are deemed a suitable fit for the number of observations (N) and large number of time-series observations (T) employed in our study. Gujarati (2004) provides prior research that explains if T is large and N is small, there is likely to be little difference in the values of the parameters estimated by the two methods and choice is made based on computational convenience. On this score, the author provides that the fixed effect method is preferred. If N is large and T is small, the estimate obtained by the two methods differs significantly.

Pooled Mean Group ("PMG") and Mean Group ("MG") Methods

We adopt the PMG and MG estimation methods for assessing robustness. Firstly, according to Bassanini and Scarpetta (2001), the advantage of a panel data for the analysis of growth equations is that country-specific effects can be controlled for by employing methods such as the Dynamic Fixed Effects and Generalized Method of Moments (GMM). However, the authors state that these approaches generally impose homogeneity of all slope coefficients, allowing only the intercepts to vary across countries. They further highlight that past research has found under slope heterogeneity these estimates become affected by a potentially serious heterogeneity bias, especially in small country samples. To this effect, the authors found that PMG was a suitable method for their sample country size of 21 countries. Due to the small country sample in our study, the PMG and MG estimators are also employed for further testing of the results.

These methods provide different treatments on the intercept, slope and error variances across the groups. The MG estimator in particular relies on estimating N time-series regressions and averaging the coefficients, whereas the PMG estimator relies on a combination of pooling and averaging of coefficients (Blackburne & Frank, 2007). In the case of MG, it allows for the intercepts, slope coefficients and error variances to differ across groups. PMG allows short-run coefficients, speed of adjustment and error variances to differ across countries but impose homogeneity on long run coefficients (Bassanini & Scarpetta, 2001).

Moreover, according to Blackburne and Frank (2007) non-stationarity tends to be a concern and the techniques suitable to estimate non-stationary dynamic panels in which the parameters are heterogeneous across groups that are the PMG and MG estimators. A unit root test is conducted to investigate the condition of stationarity of the variables, for example, if they are stationary after first differencing. The PMG and MG estimators are appropriate whether the variables are non-stationary or stationary or are a mixture of the two. In this regard, the PMG and MG methods address the non-stationary issue that may arise and are useful methods to adopt in order to further support our results. In deciding which of the PMG and MG methods to adopt, the long-run homogeneity hypothesis will be tested using the Hausman Test.

The Results

Recalling that the empirical models here represent the following: Model 1 results are derived from testing the presence of the resource curse through the resource growth equation. Model 2 results determine whether the institutions variables have any role in resolving the resource curse hypothesis. The results of Model 3 provide explanations on whether resource abundance influences the quality of institutions. For each model, we present the results in the accompanying tables, first reporting the baseline specification. We then added or removed additional determinants to illustrate how the coefficient of interest, oil rent, was affected by these alterations in the specification.

Empirical Results for Model 1—Resource Growth Equation

A negative and significant relationship between oil rent and GDP indicates the presence of the resource curse. On the basis of the Hausman Test, we find that the fixed effect model is preferred. Table 3.2 presents the results from the fixed effect estimator of Model 1.

Table 3.2: Panel Estimation Results of Model 1 (Dependent Variable GDP).

Independent Variable	Regression (1) With FD	Regression (2) With M2	Regression (3) With FD and Trade w/out M2	Regression (4) With M2 and Trade w/out FD
OilRent	–0.009	–0.062	–0.054	–0.096
	(–0.17)	(–1.14)	(–0.92)	(–1.64)
Inv	0.228**	0.284***	0.238**	0.294***
	(2.09)	(2.62)	(2.20)	(2.71)
Gce	–0.620***	–0.586***	–0.553***	–0.534***
	(–4.93)	(–4.64)	(–4.28)	(–4.09)
Inst	0.059**	0.072***	0.048*	0.065**
	(2.24)	(2.76)	(1.82)	(2.43)
Credit	0.318***		0.321***	
	(5.10)		(5.18)	
M2		0.635***		0.618***
		(4.49)		(4.36)
Trade			0.300**	0.228
			(2.01)	(1.50)
R-squared	0.88	0.88	0.88	0.88
_cons	3.273	1.868	2.034	0.991
	(5.29)	(2.45)	(2.33)	(1.03)
N	290	290	290	290

Notes: For this sample, the Hausman Test[43]favors the fixed effect method, therefore all models are estimated using the fixed effect method. *t* statistics in parentheses* *, **,*** *indicate significance at 10%, 5% and 1% respectively*.

4. The Hausman Test generated a zero p-value, which indicates the results of the two models are significantly different from each other. This is also an indication that the assumption required for the validity of the random effects model is violated (Glenn, 2010).

The specification in regression (1) refers to the baseline model with core variables. We alter the specification of the model in regression (2) by replacing credit variable with money supply as an alternative measure for financial development. In regressions (3) and (4), we include the trade variable and alternate between using the credit and money supply variable. The t-statistic is reported in parenthesis. Regressions (1) to (4) provide a negative sign for the oil-rent variable; however, it is not statistically significant and cannot be relied on to confirm the resource-curse theory. We can see from the results that almost all other variables are statistically significant. The investment variable is positive and significant at the 1% and 5% levels across all the regressions. Government expenditure indicates an inverse relationship with GDP and is consistently significant at the 1% level. The institutions variable is significant across all regressions at least at the 10% significant level. Across all regressions, the institutions variable has a positive sign indicating that an increase in the quality of institutions leads to an increase in GDP.

The credit variable was also found to have a positive effect on GDP and is significant at the 1% level, thereby indicating that it plays a role in affecting economic growth. The same can be seen for the money supply variable in regressions (2) and (4) when we alternate it with the credit variable as another measure of financial development. The trade variable is positive, however, it is only significant in regression (4) when tested with the money supply variable and does not become significant when the money supply variable is replaced with credit. In summary, the results indicate that while oil rent has the expected sign according to the resource-curse theory, it is not statistically significant. Thus, the resource-curse hypothesis cannot be established at this stage. Nonetheless, we find that almost all of the other determinants are statistically significant, which means that they are important determinants and play a role in affecting economic growth.

We find through the oil-rent variable in Model 1 no evidence of the resource curse in these countries. However, it is still too preliminary to infer from these results that resource curse does not exist for at least two reasons. First, the panel data framework pools the countries' effects and aggregates its coefficients together, and as a result factors that contribute to the presence of the resource curse may cancel each other out. For example, we see from our sample of countries that there is a varied degree of institutional levels. At the lowest end, the level of institutions for Nigeria is 3.04 and at the higher end for Malaysia it is 4.5. In this regard, the resource curse affecting countries with low levels of institutions may be neutralized with countries with higher institutional quality when they are pooled together. As a result, the outcome presented at this stage results in insignificant coefficients for the oil-rent variable. Thus, we explore whether there are conditional factors that influence the relation between oil rent and output. More specifically, we look at whether the level of institutional quality may influence how oil rent affects GDP. Secondly, we believe that oil rent could also indirectly affect economic growth through its impact on institutions.

Nonetheless the results of Model 1 present us with variables that are important determinants of economic growth for the countries in the study. These variables are those that are found to be consistently significant across all (4) regressions. We found investments are an essential input for growth as posited by the neoclassical growth model. The same is noted for the credit and money supply variables as measures for financial development. We found a significant and positive relationship between financial development and GDP. Our results are in line with Yuxiang and Chen (2011), Levine (2000) and Lucas (1988) who found that financial development has a positive impact on economic growth. This finding counters the results found by Samargandi, Fidrmuc, and Ghosh (2013) and Barajas, Chami, and Yousefi (2013). Barajas et al. found that financial development commonly has a lower, if not negative, effect on economic growth in oil rich and Middle Eastern and North African countries.

The authors further explain that this finding could indicate the financial sector was still relatively underdeveloped, and below a certain threshold, beyond which it would be capable of promoting economic growth. We observe that our result is more in line with the research of Latif et al. (2014) and Lucas (1988) who found financial development as driving economic growth. Latif et al. postulates that financial development spurs economic growth by increasing capitalization. Lucas highlights that the financial sector has a positive role and bolsters growth through mobilizing savings, allocating resources towards productive investments, reducing transaction costs and facilitating the exchange of goods and services.

The same can be seen for the institutions variable. Similar to earlier studies, such as Knack and Keefer (1995, 2012), Baltagi et al. (2009), Brunnschweiler (2008) and Glaeser et al. (2004), our results in regressions (1) to (4) found institutions to be an important determinant of economic growth. According to El Anshasy and Katsaiti (2013) the notion of institutions as a deep determinant of growth has been growing for more than a decade. They further explain that the positive relation between institutions and growth can be seen when the governing systems that have put in place support a strong rule of law enforcement and provide sound conflict management resolutions. Knack and Keefer conclude that institutions that protect property rights are crucial to economic growth and investment by providing evidence that institutions are the main determinant of income. We have also found that the government expenditure variable is inversely related with economic growth.

In this regard, our results are more in line with Solow's crowding-out theory as explained by Ahmed and Miller (1999), who described government expenditure as crowding out private investment because of higher demand for goods and services, which, in turn, results in an increase in interest rates and makes capital more expensive. We observe that in the resource-curse literature, El Anshasy and Katsaiti (2013) reported the same negative relationship between government expenditure and GDP. However, our results found government expenditure to be significant whereas the work of El Anshasy and Katsaiti did not. The authors also cite earlier literature that found a negative effect of government expenditure, which, in turn, was exacerbated by poor quality institutions. Although this linkage is not tested in our study, our findings do confirm that government expenditure and economic growth are inversely related.

Empirical Results for Model 2—Resource Growth Equation with Interaction Term

In Model 2, we introduce the interaction term into the resource growth equation to determine whether institutions have any role in alleviating the resource curse in these countries. The results are presented in Table 3.3.

Table 3.3: Panel Estimation Results of Model 2 (Dependent Variable GDP).

Independent Variable	Regression (1) With Inst*OilRent
OilRent	–0.761***
	(–3.05)
Inv	0.282***
	(2.61)
Gce	–0.475***
	(–3.64)
Inst	–0.507***
	(–2.63)

Independent Variable	Regression (1) With Inst*OilRent
FD	0.310***
	(5.06)
Trade	0.291*
	(1.97)
Inst*Oil Rent	0.190***
	(2.91)
R-squared	0.88
_cons	3.787
	(3.61)
N	290

Notes: For this sample, a Hausman Test[54]favors the fixed effect method therefore all models are estimated using fixed effect method. t statistics in parentheses[*] *, *** *indicate significance at 10% and 1% respectively.*

We start to see in Table 3.3 the signs of the resource curse as oil rent becomes negatively correlated with GDP. Thus, according to our theory and prediction β_1 is negative. The oil rent variable is also consistently significant at the 1% level. Based on the coefficient for oil rent in regression (1), a 1% increase in oil rent is associated with a 0.76% decrease in the level of GDP. Next, we turn to the oil rent and institutions interaction term. We found that not only is the interaction term positive, but it is also statistically significant at the 1% level, suggesting that the impact of institutions and oil rent on economic growth must be viewed together as an interactive effect. The positive sign of the interactive term tells us that oil rent is less likely to have a positive effect on economic growth when the quality of institutions is weak. In this regard, we can tabulate the growth impact of a marginal increase in oil rent implied by regression (1) as:

$$\frac{d\,(\text{Growth})}{d\,(\text{resource abundance})} = -0.76 + 0.190(INST)$$

Equation iii

5. The Hausman Test generated a zero p-value, which indicates the results of the two models are significantly different from each other. This is also an indication that the assumption required for the validity of the random effects model is violated (Glenn, 2010).

Based on Equation iii, we are able to solve for the institutional quality threshold as follows:

$$\frac{0.76}{0.190} = 4$$

We can infer from the results that resource curse is stronger if the level of institutional quality is low. More specifically, if the level of institutional quality is below the threshold 4, resource curse becomes stronger. On the other hand, resource curse does not apply to countries with levels of institutional quality higher than 4 as they have sufficient institutional quality to neutralize the negative effect of the resource curse. We calculate the impact of oil rent on GDP for the average, lowest and highest levels of institutional quality in Table 3.4 below.

Table 3.4: Marginal Effects of Oil Rent on GDP.

	Average *All countries*	Lowest *Nigeria*	Highest *Malaysia*
INST	3.68	3.04	4.5
$\frac{\Delta GDP}{\Delta OilRent}$	–0.062	–0.183	0.094

For the marginal effects of oil rent and institutions, Table 3.4 indicates that on average for all countries in the study the level of institutions is at 3.68. At this level, a 1% increase in oil rent is associated with a 0.06% reduction in GDP. Collectively, and as an aggregate the results show us that resource abundance is detrimental to GDP, thereby indicating a presence of the resource curse. In the second, column we tabulate the effect of oil rent on GDP for Nigeria, the country that indicates the lowest level of institutional quality. We can see that the resource curse effect is stronger as a 1% increase in oil rent results in a 0.18% reduction in GDP.

The impact of the curse is neutralized if the level of institutional quality exceeds the threshold amount. We confirm this by assessing the results for Malaysia, the country with the highest level of institutional quality in our sample of OIC countries and find that at the highest level of institutional quality, a 1% increase in oil rent contributes positively to GDP by 0.09%.

The results found in Table 3.3 confirm the findings in the literature on resource curse theory (see Alexeev & Conrad, 2011; Bjorvatn et al., 2012;

El Anshasy & Katsaiti, 2013; Latif et al., 2014; Sachs & Warner, 2001). These studies have found the oil rent coefficient to be inversely linked with the dependent variable GDP indicating that oil rent is negatively related to economic growth. The finding validates the resource-curse theory and strengthens the hypothesis that oil resources are a curse rather than a blessing for the countries in the study. But this result is because of the ineffective institutions as we have postulated and confirmed in our study.

According to Sachs and Warner (2001) a resource curse finding runs against common discussion of growth advantages. Sachs and Warner further stressed that natural resources should bring beneficial effects to development efforts and in the presence of the resource curse the common expectation of the natural resource benefits is diminished. Although the resource curse was not directly present in Table 3.2, the results in Table 3.3 tell us that with the introduction of the interaction term, the resource curse is present through another mechanism. In this respect, based on further exploration of the resource-curse hypothesis we find evidence that the resource curse only appears in countries with inferior institutions. In other words, the effects of oil rent on GDP are conditional on the levels of institutions.

We observe that Bjorvatn et al. (2012), Mehlum et al. (2006) and Bulte et al. (2005) document similar findings in their study of the oil rent and institutions interaction term with economic growth. Mehlum et al. (2006) contends that quality of institutions determines whether countries avoid the resource or not because the combination of resource abundance and weak institutions leads to low growth. Countries with higher levels of institutional quality help countries to take full advantage of their natural resources. Bjorvatn et al. (2012) cautions increasing oil rents in a situation of weak institutions due to the detrimental effects it causes on economic growth. The authors also found that the positive effect of oil rent on economic growth is pronounced when levels of institutions are strengthened. Thus, our results fare well with the literature, which have shown that natural resource abundance is particularly harmful for growth in countries with weak institutions.

Empirical Results for Model 3—Institutions and Oil Rent Relations

In Model 3, we continue to explore the impact of institutions and oil rent in order to determine whether oil rent affects GDP indirectly through its impact on institutions. Table 3.5 below reports the results.

Table 3.5: Estimation Results of Model 3 (Dependent Variable Institutions).

Independent Variable	Regression (1) Baseline	Regression (2) With Un	Regression (3) With TE	Regression (4) With Pops
OilRent	–0.733***	–0.743***	–0.764***	–0.768***
	(–5.89)	(–5.98)	(–6.23)	(–6.24)
GDP	0.396***	0.377***	0.126	0.077
	(3.13)	(2.98)	(0.83)	(0.43)
Inv	0.035	–0.025	0.155	0.170
	(0.14)	(–0.10)	(0.61)	(0.66)
Trade	1.190***	1.157***	0.946***	0.940***
	(3.64)	(3.54)	(2.86)	(2.84)
Unemployment		–0.439	–0.291	–0.290
		(–1.57)	(–1.04)	(–1.03)
Tertiary education			0.361***	0.318**
			(2.89)	(2.12)
Population				0.246
				(0.51)
R-squared	0.52	0.52	0.54	0.54
Hausman test	0.0061	0.0035	0.0015	0.0033
_cons	–0.142	1.527	1.145	–0.371
	(–0.10)	(0.88)	(0.67)	(–1.1)
N	290	290	290	290

Notes: For this sample, the Hausman Test[6]5favors the fixed effect method therefore all models are estimated using the fixed effect method. *t* statistics in parentheses* **, *** indicate significance at 5% and 1% respectively.

There are several noteworthy results that can be derived from Table 3.5. The first finding is that from regression (1) onwards oil rent has a negative relationship with institutions, suggesting that oil rent affects institutions negatively. The coefficients are also significant at the 1% level for regressions (1) to (4). As an example, in regression (4), a 1% increase in oil rent would result in a 0.77% decrease in quality of institutions. We can see that quantitatively

6. The Hausman Test generated *p-value*<0.05 is taken as evidence that fixed effects and random effects are different enough to reject the null hypothesis and hence to reject the random effects method in favor of the fixed effect method (Clark & Linzer, 2012).

the adverse impact of oil rent and institutions increases across regressions (1) to (4) with the inclusion of the control variables. The coefficient that is consistently significant at the 1% level result is the trade variable, with a positive relationship with institutions, indicating that a 1% increase in trade openness will lead to an increase in quality of institutions.

This result is in line with Rodrik, Subramanian, and Trebbi (2004) who found that trade and institutions exert a positive impact on each other. The effects of all other control variables on institutions are also as expected. We find the results of Model 3 to be in line with the work of Costa and Santos (2013), Samargandi et al. (2013), Bjorvatn et al. (2012), Alexeev and Conrad (2011), Bulte et al. (2005) and Torvik (2002). Torvik's suggested theory to explain the inverse relationship between oil rent with quality of institutions centers on the notion that natural resource rent diverts entrepreneurial talent from productive activities to those that are low in productivity, such as rent-seeking activities and concludes that the fall of income due to this resource re-allocation of entrepreneurs may outweigh the natural resource benefits of natural resource rents.

Other authors, such as Costa and Santos (2013) and Mehlum et al. (2006), also base the model for weakness in institutions as premise for the resource-curse phenomenon. Our finding is in line with the strand of studies that sees the resource-curse phenomenon involving institutional integrity. For example in the study of Alexeev and Conrad (2011), the authors examined the relationship between resource abundance and quality of institutions. Our results are also in line with theirs in respect of the negative correlation found between oil rent and quality of institutions. We can interpret from our results that resource abundance has not contributed to improved institutional quality in the countries under study when in principle one normally expects wealth generated by natural resources would contribute to the development of localities (Costa & Santos, 2013). It instead affects institutional quality adversely.

According to David and Tilton (2005), the deterioration in institutional quality at worse leads to insurrection and war and at best results in incompetence and poor planning. Similarly Sala-i-Martin and Subramanian (2003) found that natural resources do indeed lead to and are intrinsic to a decline in institutional quality. The authors believe that where natural resources are an endowment, countries that do have them are destined to institutional decline and poor growth. In a country specific analysis, Sala-i-Martin and Subramanian show that the main problem affecting the Nigerian economy was that oil revenues that the government gets are regarded as manna from heaven, which tends to undermine the quality of institutions. These results are intuitive from our earlier discussion. In the presence of large oil rents, rent-seeking and corrupt activities are encouraged. If institutions are weak, those in power have no incentive to improve institutional quality that would, in turn, reduce their ill-gotten benefits.

As for the relationship between trade and institutions, earlier literature postulates that as trade openness increases, governments face outside pressure to increase the level of institutional quality. In summary, our findings in this model show that oil rents have a negative sign, implying that oil resources are detrimental to institutional quality. Since the significant relationship between oil rent and growth has been established in Model 1, the results in this model imply that the resource curse also applies to economic growth outcomes indirectly through the institution channel. Our results further strengthen the argument that directing efforts to improve quality of institutions may be the most effective channel for ameliorating the negative impact of the resource curse.

Empirical Results for Model 4—Financial Development and Oil Rent Relations

In Model 4, we explore the impact of financial development and oil rent in order to determine whether oil rent affects GDP indirectly through its impact on financial development. Table 3.6 below reports the results.

Table 3.6: Estimation Results of Model 4 (Dependent Variable Financial Development).

Independent Variable	Regression (1) Baseline	Regression (2) With Trade	Regression (3) With Foreign*Trade
OilRent	**–0.230*** **	**–0.234*** **	**–0.212*** **
	(–4.64)	(–5.98)	(–6.23)
Inv	0.291**	0.293**	0.358***
	(2.81)	(2.83)	(3.27)
Foreignlag3	0.002	0.000	0.006
	(0.17)	(0.01)	(0.40)
Tradelag3		0.062	0.128
		(0.47)	(0.93)
Foreign*Trade			–0.007*
			(–1.79)
R-squared	0.75	0.75	0.75

Independent Variable	Regression (1) Baseline	Regression (2) With Trade	Regression (3) With Foreign*Trade
_cons	3.145	2.931	2.586
	(8.91)	(5.06)	(4.25)
N	290	290	290

Notes: For this sample, the Hausman Test[76]favors the fixed effect method therefore all models are estimated using the fixed effect method. *t* statistics in parentheses* *, **,*** *indicate significance at 10%, 5% and 1% respectively.*

The results in Table 3.6 point to a negative relationship between oil rent and financial development. The coefficient of the oil rent variable is also consistently significant and across regressions (1) to (3) for all alternative specifications. We can see in regression (2) that the degree of the negative relationship is strengthened with the inclusion of the trade variable. However, in regression (3), the impact becomes less with the inclusion of the interaction term between foreign investment and trade. We find that a 1% increase in oil rent is associated with a 21% to 23% reduction in level of financial development across the regressions. The investment variable is positive for the regressions and consistently significant at the 1% level while other variables are generally not significant.

Our model was based on the specification by Yuxiang and Chen (2011), and thus we were able to generate the same results between oil rent and financial development. Similar to Yuxiang and Chen, we found a negative and significant relationship between resource abundance and financial development, thereby indicating that resource abundance adversely impacts financial development. Having earlier established the positive and significant linkage between financial development and growth, Yuxiang and Chen (2011) were also able to conclude that due to the resource curse, financial development deteriorates and as a result economic growth becomes adversely affected. We followed the same strategy by establishing the same linkage between financial development and growth in Model 1. Our results indicate financial development to be an important determinant of economic growth. As a result of the negative relationship between resource abundance and financial development and because we found that resource abundance is detrimental to financial

7. The Hausman Test generated *p-value*<0.05 is taken as evidence that fixed effects and random effects are different enough to reject the null hypothesis and hence to reject the random effects method in favor of the fixed effect method (Clark & Linzer, 2012).

development, we take the same view to infer that resource curse affects economic growth indirectly through the financial development channel.

We refer to the work by Latif et al. (2014) and Van der Ploeg (2009) to derive plausible explanations for this relationship. According to Latif et al., investments are impacted due to misallocation of resources, which is triggered by corruption or rent-seeking activities. The authors go on to add that because financial development plays a vital role in stimulating domestic output by enhancing investment opportunities, the effects of the resource curse cause impairment to financial development and as a result economic growth becomes adversely affected.

Van der Ploeg further hypothesises that due to the high volatility, resource revenues become highly volatile because their supply exhibits a low price elasticity. As a result, fluctuation in exchange rate ensues, and countries with low levels of financial development experience adverse impact on long-term productivity growth. Van der Ploeg cites evidence that show the adverse growth effect of natural resources results mainly from volatility of commodity prices, especially for economies with weak financial institutions, current account restrictions and high capital account mobility. Furthermore, due to bankruptcy costs, relative price volatility triggers an increase in interest rate. According to Van der Ploeg volatility is closely related to the resource curse and through financial development the effects of the curse may be offset. We conduct robust analysis below, to confirm the results we have obtained thus far.

Robustness Analysis

In the final step of our analysis, we assess whether we are able to generate the same results for Models 1 to 4 using the PMG method. Additionally, we also conduct the analysis for non-OIC top oil-producing countries[87]with reported sound institutional structure[98]to confirm that the resource curse is not a factor in these countries.

PMG Method

As previously discussed, the PMG method is suitable to estimate non-stationary dynamic panels in which the parameters are heterogeneous across the group and address non-stationarity concerns. Table 3.7 reports the results.

8. Oil-exporting countries listed by US Energy Information Administration.
9. Top institutional quality analysis by ICRG.

Table 3.7: PMG Estimation Results for Models 1–4.

Independent Variable	Regression (1) Baseline D.GDP	Regression (2) With Inst*OilRent	Regression (3) D. Inst	Regression (4) D.FD
OilRent	**–1.141***	**–1.113**	**–0.504***	**–0.102**
	(–3.43)	(–0.93)	(–4.11)	(–0.61)
Inv	1.887***	1.402*	0.323	2.634***
	(3.69)	(2.42)	(0.10)	(6.19)
GCE	3.101***	1.595*		
	(3.37)	(2.19)		
Inst	0.186*	–1.165		
	(2.44)	(–1.20)		
FD	–0.369	–1.160*		
	(–1.68)	(–2.11)		
Trade		–1.243	1.824***	
		(–1.37)	(3.39)	
Inst*OilRent		0.329		
		(1.00)		
GDP			–0.550	
			(–1.74)	
Unemployment			–0.560	
			(–1.44)	
Tertiaryeducation			0.253	
			(1.01)	
Population				
Population			0.920	
			(1.13)	
Foreignlag3				0.810
				(0.87)
Tradelag3				1.958
				(0.90)
Foreign*Trade				–0.241
				(–0.97)
N	280	280	280	277

Regression (1) depicts the test results for Model 1. We find that resource curse is present through the significant and negative relationship between oil rent and GDP. In this regard, oil rent has a direct negative effect on GDP. We find that with the exception of financial development, all other variables in regression (1) are significant. For example, we can see that institutions are significant at the 10% level, which reaffirms our prediction that institutions are an important determinant of economic growth.

Regressions (2) and (3) provide results for Model 1 in the study of the interaction terms between oil rent and institutions. In contrast with our findings in Table 3.3, although oil rent is negatively signed, we were not able to generate statistically significant results for oil rent as a standalone variable and when it is interacted with institutions. Nonetheless, the sign of the interaction term between oil rent and institutions is in accordance to our theory as it shows a positive relationship with GDP.

Finally, we were able to generate a negative sign for the oil rent variable in Model 4, seen in Table 3.7 as regression (5), but we were not able to obtain a statistically significant outcome. The results thus indicate that there is not enough evidence to justify that financial development is a useful channel and mechanism to reduce the adverse effects of the resource curse. Although we were not able to generate exact outcomes for the models between the fixed effect and PMG methods, there are important similarities that we can note from the results. Firstly, estimation by both methods showed the presence of the resource curse. According to the fixed effect method, the resource curse is present through various other mechanisms such as indirectly through institutions. In the PMG method, we were able to note the existence of the resource curse directly through the oil rent relationship with GDP in Model 1 and indirectly through the institutions channel through the oil-rent relationship with institutions. As such, traces of the resource curse are evident by both methods and the negative effect of resource abundance on GDP whether directly or indirectly cannot be dismissed. Secondly, both methods indicate that institutions as a standalone variable are an important contributor to economic growth and that resource abundance causes deterioration to the quality of institutions. In this regard, both methods support our prediction that institutions are a channel that account for the presence of the resource curse thus focusing efforts on improving the quality of institutions may prove to be essential to generate a positive and indirect effect to increase economic growth.

We were not able to prove in the PMG method that financial development is an important mechanism to resolve the resource curse. We also did not find in the PMG method that financial development is an important contributor to economic growth as the results were insignificant. In this regard, through the estimation of both methods, financial development is found to be inferior to institutions in its relation with oil rent and GDP.

We thus reach the following conclusions:

1. Whether directly through the relationship between oil rent itself with GDP or indirectly through the impacts of oil rent with other channels, we cannot dismiss the assertion that the resource curse is present for all of the countries in the study.
2. There is strong evidence based on the consistent results obtained using both methods to show that natural resource abundance is inversely related to the quality of institutions. We agree with the literature that demonstrates resource abundance induces counterproductive behaviors such as corruption and rent-seeking activities and that improving the quality of institutions would ameliorate the negative fallouts of the resource curse.
3. We were not able to show financial development as an important determinant of economic growth; thus the results are in line with our prediction that the impact of oil rent on GDP is not conditional on the level of financial development but is significantly dependent on the quality of institutions.

Comparison with Top Oil-Producing Non-OIC Countries

In theory, we would predict that the resource curse is not present in countries with sound institutional quality. We ran the base equation (Model 1) for selected top 25 non-OIC oil-producing countries[109]to confirm that the resource curse was not present in these countries. Results are reported in Table 3.8 next page.

The results indicate that oil rent is significant and positively related with GDP. A 1% increase in oil rent is associated with a 0.22% increase in levels of GDP, thereby confirming that oil rent for these countries does not have a detrimental effect to the country's GDP and its growth. Oil rent, thus represents

10. The countries selected are Australia, Canada, Norway and United Kingdom.

Table 3.8: Panel Estimation Results of Model 1.

Independent Variable	Model 1 (1)
OilRent	0.225***
	(4.24)
Inv	0.184
	(0.70)
Gce	0.632
	(1.24)
Inst	0.012
	(0.39)
Credit	1.695***
	(12.47)
M2	–0.348**
	(–2.06)
Trade	–0.437
	(–1.49)
R-squared	0.81
_cons	–2.57
	(–1.79)
N	116

Notes: For this sample, the Hausman Test[11] favors the fixed effect method therefore all models are estimated using fixed effect method. *t* statistics in parentheses* **, *** *indicate significance at 5% and 1% respectively.*

more a blessing and not a curse as in the case of our OIC countries, which are burdened by ineffective institutions. These results confirm our findings under Model 2 and support our argument that countries with higher levels of institutional quality are spared from the negative fallouts of the resource curse and instead benefit from their natural resource abundance.[10]

11. The Hausman Test generated *p-value*<0.05 is taken as evidence that fixed effects and random effects are different enough to reject the null hypothesis and hence to reject the random effects method in favor of the fixed effect method (Clark & Linzer, 2012).

Conclusion

Our core findings in this chapter suggest firstly that resource curse is present in the countries under study, whether directly through the oil-rent variable or indirectly through the relationship of oil rent with other channels. The detrimental effects of a resource curse cannot be ignored and further analysis was conducted to determine its cause. Our findings are consistent with the consensus that has appeared around the landmark study of Sachs and Warner and the emerging studies in the resource curse that have followed. Our results in Models 2 and 3 indicate that the association between oil rent and growth operates to a significant extent through the channel of institutional quality.

The significant relationship indicates that weakness in the quality of institutions is important in how oil rent affects economic growth. If institutions are effective and of high quality, then oil rents, namely, an abundance of natural resources, support economic growth. But if institutional quality is low, we can expect the abundance of natural resources to impair economic growth. When we explored the oil rent and institutions relation further and treated institutions as a dependent variable, we found that oil resources had a detrimental effect on the quality of institutions; this should be expected as those in power have the incentive to impede institutional improvement so that they can continue to benefit from resource rents.

When we compared financial development and institutions improvement, we found that it is more productive to focus on the latter as the channel to eliminate the adverse fallout of the curse. Our findings surrounding the resource curse and the role of institutions as the important channel to ameliorate the fallout and to turn the curse into a blessing will drive the policy implications and our recommendations in the following chapters of this book.

References

Ahmed, H., & Miller, S. M. (1999). Crowding-out and crowding-in effects of the components of government expenditure department of economics working paper series. *Economics Working Papers*, *199902*.

Alexeev, M., & Conrad, R. (2011). The natural resource curse and economic transition. *Economic Systems*, 35(4), 445–461. https://doi.org/10.1016/j.ecosys.2010.10.003

Asteriou, D., & Hall, S. G. (2007). *Applied econometrics: A modern approach*. Hampshire: Palgrave Macmillan.

Auty, R. M. (2007). Natural resources, capital accumulation and the resource curse. *Ecological Economics*, *61*, 627–634. https://doi.org/10.1016/j.ecolecon.2006.09.006

Ayelazuno, J. (2014). Oil wealth and the well-being of the subaltern classes in Sub-Saharan Africa: A critical analysis of the resource curse in Ghana. *Resources Policy*, *40*, 66–73. https://doi.org/10.1016/j.resourpol.2013.06.009

Clark, T. S., & Linzer, D. a. (2012). Should I Use Fixed or Random Effects? *Mimeo*, 1–34. https://doi.org/10.1017/psrm.2014.32

Baltagi, B. H., Demetriades, P. O., & Law, S. H. (2009). Financial development and openness: Evidence from panel data. *Journal of Development Economics*, 89(2), 285–296. https://doi.org/10.1016/j.jdeveco.2008.06.006

Barajas, A., Chami, R., & Yousefi, S. R. (2013). The finance and growth nexus re-examined: Do all countries benefit equally? *IMF Working Paper*, *13/30*.

Bassanini, A., & Scarpetta, S. (2001). Does human capital matter for growth in OECD countries? Evidence from pooled mean-group estimates. *OECD Economic Department Working Papers*, *74*, 399–405. https://doi.org/10.1787/424300244276

Bjorvatn, K., Farzanegan, M. R., & Schneider, F. (2012). Resource curse and power balance: Evidence from oil-rich countries. *World Development*, *40*(7), 1308–1316. https://doi.org/10.1016/j.worlddev.2012.03.003

Blackburne, E. F., & Frank, M. (2007). Estimation of nonstationary heterogenous panels. *The Stata Journal*, *7*(2), 197–208.

Boschini, A., Pettersson, J., & Roine, J. (2013). The resource curse and its potential reversal. *World Development*, *43*, 19–41. https://doi.org/10.1016/j.worlddev.2012.10.007

Brooks, C. (2008). *Introductory econometrics for finance* (2nd ed.). https://doi.org/10.1111/1468-0297.13911

Brunnschweiler, C. N. (2008). Cursing the blessings? Natural resource abundance, institutions, and economic growth. *World Development*, *36*(3), 399–419. https://doi.org/10.1016/j.worlddev.2007.03.004

Brunnschweiler, C. N., & Bulte, E. H. (2008). The resource curse revisited and revised: A tale of paradoxes and red herrings. *Journal of Environmental Economics and Management*, *55*(3), 248–264. https://doi.org/10.1016/j.jeem.2007.08.004

Bulte, E. H., Damania, R., & Deacon, R. T. (2005). Resource intensity, institutions, and development. *World Development*, *33*(7), 1029–1044. https://doi.org/10.1016/j.worlddev.2005.04.004

Costa, H. K. D. M., & Santos, E. M. Dos. (2013). Institutional analysis and the "resource curse" in developing countries. *Energy Policy*, *63*, 788–795. https://doi.org/10.1016/j.enpol.2013.08.060

David, G. A., & Tilton, J. E. (2005). The resource curse. *Natural Resources Forum*, *29*, 233–242. https://doi.org/10.1080/05679320500129037

Edwards, S. (1992). Trade orientation, distortions and growth in developing countries. *Journal of Development Economics*, *39*(1), 31–57. https://doi.org/10.1016/0304-3878(92)90056-F

El Anshasy, A. a., & Katsaiti, M.-S. (2013). Natural resources and fiscal performance: Does good governance matter? *Journal of Macroeconomics*, *37*, 285–298. https://doi.org/10.1016/j.jmacro.2013.05.006

Glaeser, E. L., La Porta, R., Lopez-De-Silanes, F., & Shleifer, A. (2004). Do institutions cause growth? *Journal of Economic Growth*, 9, 271–303.

Glenn. (2010). Eviews.Com. Retrieved from http://forums.eviews.com/viewtopic.php?f=18&p=7732

Gujarati, D. N. (2004). *Basic econometrics. Basic econometrics* (Vol. 4.). https://doi.org/10.1057/9780230226203.0425

James, A., & Aadland, D. (2011). The curse of natural resources: An empirical investigation of U.S. counties. *Resource and Energy Economics*, 33(2), 440–453. https://doi.org/10.1016/j.reseneeco.2010.05.006

Knack, S., Keefer, P., 1995. Institutions and economic performance: cross-country tests using alternative institutional measures. Econ. Polit. 207–227.

Knack, S., & Keefer, P. (2012). Institutions and Economic Performance: Cross-Country Tests Using Alternative Institutions Indicators. *MPRA Paper World Bank*, *23118*, 2.

Kolstad, I., & Wiig, A. (2012). Testing the pearl hypothesis: Natural resources and trust. *Resources Policy*, *37*(3), 358–367. https://doi.org/10.1016/j.resourpol.2012.04.003

Larsen, E. R. (2005). Are rich countries immune to the resource curse? Evidence from Norway's management of its oil riches. *Resources Policy*, 30(2), 75–86. https://doi.org/10.1016/j.resourpol.2004.12.001

Latif, S., Farooq, A., Loganathan, N., & Shahbaz, M. (2014). Empirical evidence on the resource curse hypothesis in oil abundant economy. *Economic Modelling*, *42*, 421–429. https://doi.org/10.1016/j.econmod.2014.07.020

Levine, R. (2000). International financial liberalization and economic growth. *Review of International Economics*, (August 18, 2000), 612–626.

Li, X., & Liu, X. (2005). Foreign direct investment and economic growth: An increasingly endogenous relationship. *World Development*, 33(3), 393–407. https://doi.org/10.1016/j.worlddev.2004.11.001

Lucas, R. E. (1988). On the mechanics of economic development. *Journal of Monetary Economics*, *22*(February), 3–42. https://doi.org/10.1016/0304-3932(88)90168-7

Mehlum, H., Moene, K., & Torvik, R. (2006). Institutions and the resource curse. *Economic Journal*, *116*(508), 1–20. https://doi.org/10.1111/j.1468-0297.2006.01045.x

Oskenbayev, Y., Yilmaz, M., & Abdulla, K. (2013). Resource concentration, institutional quality and the natural resource curse. *Economic Systems*, *37*(2), 254–270. https://doi.org/10.1016/j.ecosys.2012.11.001

Prescott, E. C. (1956). Robert M. Solow's neoclassical growth model: An influential contribution to economics. *The Scandinavian Journal of Economics*, 90(1), 7. https://doi.org/10.2307/3440145

Rodrik, D., Subramanian, A., & Trebbi, F. (2004). Institutions rule: The primacy of institutions over geography and integration in economic development. *Journal of Economic Growth*, 9(2), 131–165. https://doi.org/10.1023/B:JOEG.0000031425.72248.85

Sachs, J. D., & Warner, A. M. (1995). Natural Resource Abundance and Economic Growth. *NBER Working Paper Series*, 3, 54. https://doi.org/10.3386/w5398

Sachs, J. D., & Warner, A. M. (1997). Natural Resource Abundance and Growth. *Centre for International Development and Harvard Institute for International Department.*

Sachs, J. D., & Warner, A. M. (2001). The curse of natural resources. *European Economic Review*, 45(4–6), 827–838. https://doi.org/10.1016/S0014-2921(01)00125-8

Sala-i-Martin, X., & Subramaniam, A. (2003). Addressing the natural resource curse: An illustration from Nigeria. *IMF Working Paper, WP/03/139*.

Samargandi, N., Fidrmuc, J., & Ghosh, S. (2013). Is the relationship between financial development and economic growth monotonic for middle income countries ? Is the Relationship between financial development and economic. *World Development*, 68(13), 66–81. https://doi.org/10.1016/j.worlddev.2014.11.010

Stijns, J. C. (2005). Natural resource abundance and economic growth revisited, *30*, 107–130. https://doi.org/10.1016/j.resourpol.2005.05.001

Torvik, R. (2002). Natural Resources, Rent-Seeking and Welfare. *Journal of Development Economics*, *67(2)*, 455–470.

Tsani, S. (2015). On the relationship between resource funds, governance and institutions: Evidence from quantile regression analysis. *Resources Policy*, *44*, 94–111. https://doi.org/10.1016/j.resourpol.2015.01.003

Van der Ploeg, F. (2009). Natural resources: Curse or blessing? *CESIFO Working Paper No. 3125*.

Yuxiang, K., & Chen, Z. (2011). Resource abundance and financial development: Evidence from China. *Resources Policy*, 36(1), 72–79. https://doi.org/10.1016/j.resourpol.2010.05.002

· 4 ·

THEORIES OF INSTITUTIONS

Introduction

Our empirical results lend strong support to the resource curse theory. Our results have indicated that oil revenues in the oil-producing countries have created a drag on their economic performance. Through further investigation we found institutional quality to be an, or possibly the, important factor accounting for the deleterious effect of the resource curse. These results are particularly discerning as the membership of these oil-producing countries of the OIC indicates that they should be the exemplars of Islamic societies. The direct consequence of practicing the rules and teachings from the divine Qur'an and Sunnah should result in practices of high institutional integrity and quality. Thus, this contradiction in the expected performance and behavior merits further understanding to ensure that priorities and interests are realigned according to the rules of Allah (swt) in order to ameliorate any negative fallout of the resource curse and manage the Almighty's gift in the way that He intended.

Nevertheless, the finding is not surprising as earlier research has emphasized that the resource curse is an institutional and economic phenomenon (see Brunnschweiler, 2008; Bulte, Damania, & Deacon, 2005; Costa & Santos 2013;

Mehlum, Moene, and Torvik, 2006; Oskenbayev, Yilmaz, & Abdulla, 2013). This earlier research suggests the need for policymakers to develop strategies for reforms that address institutional weaknesses to improve economic growth performance. On this note, the discussions put forward in this chapter will first center on the definition for institutions according to the conventional framework along with the suggestions and recommendations for institutional reforms provided in this literature. We then juxtapose this definition of institutions and recommendations against the Islamic definition of institutions and suggested institutional framework in order to highlight the fundamental differences between these two worldviews.

Definition of Institutions in the Conventional Literature

According to Glaeser, La Porta, Lopez-De-Silanes, and Shleifer (2004), Douglass North's (1981) definition of institutions is the one most often followed in earlier research: that it is a set of rules, compliance, procedures, and moral and ethical behavioral norms designed to constrain the behavior of individuals. All of which, are aimed towards pursuing the interests of maximizing wealth or utility or principles. In North (1990), institutions are broadly referred to as the "rules of the game" in a society or the constraints that control human interaction. These rules are said to reduce uncertainty and provide structure to one's life. According to North, as human beings create institutions, institutions may evolve as a result of human alteration. Thus, the prerequisite to understanding the theory of institutions is to begin with an understanding of individual and human behavior.

North further explains that the major role of institutions in a society is to reduce uncertainty through establishing a structure that is stable for human interaction. This structure, however, may not necessarily be efficient. Institutions form conventions, codes of conduct, norms of behavior to statute common law and contractual obligations of individuals, and because they are said to evolve in nature, the choices made available for individuals also become continually altered (North, 1981). Within the ambit of economic institutions, the same definition and objective apply to economic management. Opportunities in society are said to be determined when institutions are coupled with the constraints of economic theory (such as scarcity and competition). According to North, organizations are created to take advantage of these opportunities and as organizations evolve, so do institutions. The resultant path of institutional change is explained as a lock-in situation arising from mutually dependent relationship between institutions and organizations. These elements have evolved as a consequence of the incentive structure provided by those institutions and by the human being feedback process pertaining to changes in the opportunity set.

Acemoglu, Johnson, and Robinson (2005) offer another insight on the origins of culture (societal norms) that shape economic development. They cite the work of Weber (1930) who had traced the origins of industrialization in Western Europe to the Protestant Reformation. In Weber's view, the intrinsic set of beliefs and the worldview of Protestantism were crucial to forming the belief system of capitalism. Protestantism emphasized the idea of predestination by stressing that some individuals are "chosen" while others are not.

Their theory expands into the notion of man and labor. Protestantism led to a set of beliefs that emphasized hard work, savings and finding means for material gains. Economic success indicates one's elite recognition by God, and as a result one may not be so willing to part with their wealth. Acemoglu et al. highlight that one of the reasons for economic differences is the differing sets of beliefs that form different religions. They put forth reasons raised by other scholars, including the adoption of cultural endowments or societal trends, that do not promote societal success or prosperity. The example given described how in some parts of the world, a culture to trust only the individuals of one's own family and to refuse cooperation or trust to others is persistently adopted. Thus, in this respect, North's suggestion to focus efforts to understand institutions by understanding human behavior merits attention. Nonetheless, Acemoglu et al. also view the rational actions of key economic and political agents in recognition of their consequences as factors that cause division between institutions.

In a later study, North (1993) likens institutions to the rules of the game and organizations as the players. Here, North sheds some light as to how the institutional change happens. He presents it as a deliberate process shaped by the perceptions of the actors where the resulting outcome is a direct consequence of their actions. The organizations he refers to are broken down to economic organizations which include firms, cooperatives, trade unions etc.; political organizations which refer to political parties; legislature and regulatory bodies; and educational organizations which include universities, schools and other centers of teaching. Although he states the immediate objectives of these organizations are to maximize profits or improve reelection prospects (for political parties), North explains that the ultimate objective is survival because all organizations live in a world of scarcity. It is against this backdrop that he describes the following five propositions for institutional change.

Institutional Change Propositions (North)

The first proposition centers on competition and how it is the key to institutional change due to continuous interactions between institutions and organizations in the economic setting of scarcity. Secondly, competition forces organizations to continually invest in new skills and knowledge to survive. The kinds of skills and knowledge that individuals and their organizations acquire will shape evolving perceptions about opportunities and choices that will incrementally alter institutions. Due to the ubiquity of competition that exists against the setting of scarcity, the key to survival is for organizations to invest in new skills and knowledge as to improve its efficiency and to beat its rivals.

Third, the institutional framework should provide the incentive structure that dictates the perceived knowledge and skills that have the maximum payoff. For example, if piracy is perceived to generate the highest rate of return in an economy, one can expect organizations to devote resources to invest in skills and knowledge that will make them better pirates. Thus, perceived benefits become the fundamental influence for their immediate investment.

The fourth proposition looked at how perceptions are derived from the mental constructs of the players. The key to choices that the players make is linked to their perception about payoffs and that each individual will carry with them different interpretations of these payoffs. As everyone's environment consists of differing informational backgrounds, they end up making different choices. In a macro view, perceptions cannot converge over time to a common equilibrium where everyone would have a correct perception due to the different mental constructs of individuals.

The fifth and final proposition relates to the viability, profitability and survival of organizations that depend on their existing institutional structure and how their current interdependent contracts and other relationships have been built into it. The two stated implications of this proposition are that institutional change is incremental and that it is path dependent. North also stresses that humans attempt to reduce uncertainty through learning. He believes that it is through cumulative learning that societies form their respective cultures, which they then scaffold into the mental models they possess and the external environments that they build—the institutions. Thus, focusing on human learning will allow one to understand the beliefs and preferences that have shaped the performance of economies through time. In his opinion, it is the learning system that induces political and economic entrepreneurs into a position that enables them to make choices. These choices, in turn, shape the micro and macro-economic performance that will erect a structure of rules, norms, conventions and beliefs embodied in governing institutions, constitutions, property rights and informal constraints. He states, that all these factors will in turn drive and shape economic performance.

The final and crucial point from North's study is that it is this belief system that has set the source of path dependence. He states, although political or economic entrepreneurs seek to alter some aspect of economic performance, they make choices that are constrained by not only the standard constraints of income and technology but also by the scaffolding of the previously mentioned mental models. As a result, it is their understanding that drives the way the economy works and the policies and framework that they construct.

Although North delineates in great detail how institutional change can take place, he is a critic of the current belief systems that have shaped the minds of the actors effecting this change. Actors that act based on incomplete information lead to unintended consequences that affect economic welfare for society at large. He opines that although the decisions they have made have increased material well-being, they have also produced death, famine, starvation, economic decline and stagnation, even the disappearance of civilizations. He concludes that the past experience that shape future institutional prescriptions do not equip members to confront and solve new problems. Changing the belief system will result in changes in institutions, organizations and economic performance.

In summary, North outlines the overarching rules of the institutions that are based on the economic conditions faced by man. However, this pursuit of survival may not always lead to efficient or just systems. In discussing the propositions of institutional change, North provides the importance of assessing the mental make-up of man and understanding the elements that drive the behavior of man. Different belief systems result in differences in institutions and economic performance. Acemoglu et al. cite previous work that expands on this notion.

In reference to North's analogy, his argument seems to point to the following: to change the game, one must understand the players and the rules they go by. North provides the flaws in belief systems that are driven by socio-cultural influences and suggests that institutional change is required. Even though institutions operate on the platform of different belief systems, the importance of understanding economic growth and performance is not to be overlooked. The form and context of later literature on institutions and economic growth are discussed in the next section.

Do Effective Institutions Support Economic Growth?

Studies devoted the nature of the relationship among institutions, good governance or political stability and economic growth have received considerable amount of attention since the mid-1990s (Sihag, 2007). Knack and Keefer (1995) are often credited for sparking this interest as they conclude that institutions which protect property rights are crucial to economic growth and investments (Sihag, 2007). Acemoglu, Johnson, and Robinson (2000) further enhanced their hypothesis and empirically proved institutions to be the main determinants of income.

In a study to further ignite this debate, Rodrik, Subramanian, and Trebbi (2004) estimate the respective contributions of institutions, geography and trade in determining income levels around the world and take the position that it is quality of institutions that matters more than everything else. Their study echoes the finding of Knack and Keefer (1995) as they found economies are richer when investors believe their property rights are protected.

The work of Acemoglu et al. (2005) shed some light as to what good institutions should look like in the economic dimension. They state that good institutions are those that provide security of property rights and fair and equal access to economic resources to all members within the society. In their view, societies face weaknesses in economic institutions when only small fractions of the population have enforced property rights. The provision of property rights should allow for individuals to have an incentive to invest, innovate, and be a part of economic activities. The authors also stress the importance of equality before the law to enable stable and just societies and for good investment opportunities to be utilized effectively.

In another strand of the literature, Rodrik et al. (2004) argue that there is a need to distinguish between economic principles and their institutional embodiment. In their view, most first-order economic principles such as incentives, competition, hard-budget constraints, fiscal sustainability and property rights do not directly map into institutional forms. In fact, they suggest some of these economic ideas to be institution free, for example, macroeconomic stability is achieved under fiscal institutions and property rights through a well-functioning regulatory authority. In this regard, the authors define their version of institutional solution as one that is made up of seamless integration between a number of supporting norms and complementary institutions.

Glaeser et al. (2004) on the other hand challenge the measures used by existing studies and claim human capital as a more accurate determinant for economic growth. Although they do not outright reject the institutional linkage with growth, they posit that institutions improve when countries accumulate human and physical capital while becoming richer in the process.

It is not our aim to continue the debate on whether good institutions cause growth or whether high growth itself affects institutional change, or even whether institutions are the *fundamental* explanation for economic growth. For a more extensive discussion on this matter, the summary of the study on institutions and economic growth by Sihag (2007) will be adopted.[1] Rather, the preceding literature provides the background for understanding institutions in light of conventional settings.

In summary, differences in institutional quality fundamentally cause different patterns in economic growth because humans across different societies (institutional setting) organize themselves differently. Secondly, in the organization of institutions and sound markets, the notion of protection of property rights (and equality before the law) becomes of primary importance to economic outcomes and even the reason for such outcomes to occur.

Institutions According to Islam

One of the most fundamental differences between the notions of institutions in Islam with that of institutions in conventional literature lies in their source. The rules governing behavior in Islam are not man made but are instead Creator centered. They are derived from the divine teachings of the Qur'an with guidance on its operationalization from Prophet Muhammad (sawa). These teachings are the words of the Supreme Creator, Allah (swt), which form a meta-framework of rules. Compliance to these rules would provide unparalleled benefits to a man's life. In contrast to the human-centered system that drives conventional forms of institutions, the rules in Islam do not change

1. According to Sihag, the matter remains inconclusive given that most of the tests conducted are specified to validate each researcher's own brand of hypothesis.

over time and are not subject to alteration. The rules remain relevant, necessary and immutable and permeate every aspect of man's life.

The fountainhead of all Islamic paradigms is the Qur'an. It provides the framework within which all relevant envisioned conceptions of reality find their source. This eternal source specifies rules of behavior (institutions) applicable to all societies at all times. These rules are immutable temporally and spatially. The economic system established by the Messenger in Medina is the Archetype of Islamic economic systems. In this Archetype, there is a core of institutional structure, which is immutable because these systems are firmly established based on the Messenger's authoritative operationalization of the rules prescribed by the Creator in the Qur'an. A typical example is the institution of inheritance where the specific procedure is described in the Qur'an on how the inheritance is to be distributed. There are also institutions which the Messenger established which, while not explicitly stated in the Qur'an, are based on the understanding of the Qur'an as its highest interpretive authority.

An example of this type of institution is the rules of market behavior. These two types of rules are immutable, i.e., any conception of how an Islamic economy works will have to take these two elements of the Archetypal Model as a given. There is a third type of institution at the periphery of the Archetypal Model that are temporally and spatially specific to the time and the place in which the Archetypal Model was implemented. For example, the Messenger instituted rules of non-interference with market forces and the need for unhindered flow of information in the market. This rule is of and itself an immutable rule of the Archetypal Model, but the forces that would interfere with market functioning may vary and are time and place dependent. For instance, before Islam one acceptable method of interfering with the market forces in Arabia was that middlemen would meet caravans bringing supplies some distance outside of the cities and purchase their supplies. The Messenger prohibited this procedure. Clearly the principle of noninterference with the market forces is unchanged, but this particular procedure is no longer relevant. The economic hermeneutics of this rule and its application to a particular time, place and market is part and parcel of what an Islamic economic paradigm would address.

The Metaframework specifies rules (institutions) that are, to a degree, abstract. The Archetypal Model articulates the operational form of these rules. The Metaframework specifies the immutable, abstract rules. The Archetypal Model demonstrates how these rules are operationalized in a human community. The abstract became operational in the hands of the one human being who was the one and only direct recipient of the source of the Metaframework, i.e., the Qur'an. Through the words and actions of this perfect human the Metaframework given by the Creator in the Qur'an was interpreted, articulated and applied to the immediate human community of his time. The Metaframework specifies general universal laws, rules of behavior. The Archetypal Model provides universal-specific rules of behavior and the institutional structure needed for organizing a human society based on the immutable rules of the Metaframework. Men are in compliance with the Creator-conceptualized concept of reality. In this reality, man distinguishes between his physical, *basar* and non-physical form *insan*, creating a difference between what is man and what is human. *Basar* refers to the physical exterior form and *insan* closely relates to internal substance of the potentialities of man. Outwardly these two forms may look the same but inwardly is what separates and controls man. The inward form determines whether man may have attributes that are worse than animals in one extreme, and on the other end of the spectrum he may be cultivating the awareness of the potentialities of the best human state that it may lead him to surpass even the angelic state.

Believers are expected take themselves to the passage of the perfect human state through continuous actualization of these potentials and degrees of compliance towards the rules of Allah (swt). Where North had earlier indicated that man's motivation is to strive against the backdrop of economic conditions, man's motivation in the Islamic model is to actualize the covenant that has been put forth by the Creator for His Creation. Understanding the very definition of the word Islam may enhance our understanding of the reasons behind actualizing this covenant.

According to Mirakhor and Hamid (2009), the words from which Islam is derived bring the connotation of "delivering oneself into security." They further define Islam as a relational activity involving acts of submitting, surrendering and delivering oneself to the safety and security of Allah (swt) and the one who engages in this activity as a "Muslim." The safety and security of Allah (swt) have the profound impact for a man to feel protected, comforted, enlightened and to experience a guarantee of life without apprehension or grief. These were mentioned in the Qur'an in Surah Yunus, Verse 62 and Surah Al-Baqarah, Verse 257:

> Allah is the Comforter of those who are mu'min (active-dynamic believers); He extracts them from all manner of darkness into light.
>
> Indeed! The devotees of Allah will have no fear upon them nor shall they grief.

The first Qur'anic verse is complete in showing the concerns and responsibilities of Muslims—they are to devote themselves to adhering to the rules of Allah (swt) during their limited time in this world. The attainment of success in the lasting life in the Hereafter becomes the uppermost goal, surpassing any material or other worldly achievements. The criterion of this success depends on staying away from anything that is repugnant in the sight of Allah (swt) and striving for righteousness.

Mirakhor and Hamid (2009) explain the fear and grief mentioned in the second Qur'anic verse as relating to the "apprehension regarding the future and grief is the result of regrets over the past." In a nutshell, in trusting Allah (swt) as the Creator, the worry towards any uncertain outcomes that man may face is vanquished. The same message is given for the idea of one's provision and sustenance according to Surah Al-Anfal, Verse 2–4, Surah An-Noor, Verse 37–38 and Surah Luqman, Verse 20 from the Qur'an.

> The true believers with complete faith are those whose hearts are filled with awe at the mention of Allah, and His attributes; and whose faith grows stronger as they listen to His revelations and they put all their trust in their Lord; They are those who observe regular prayers and who give in charity a part of what We bestowed on them as sustenance; Such are the true believers; for them are grades of honor with their Lord, and forgiveness, and a generous sustenance in Paradise.

> The worship of Allah is performed by people of firm faith whom neither trade nor business dealings can divert them from remembering Allah, from performing prayers, or from giving to the poor, they are those who dread the day of judgement; that Allah may give them the best reward of what they have done, and give them more out of His grace; and Allah gives sustenance to whom He pleases without measure.

> Do you not see that Allah has subjected to you (use) all things in the heavens and on earth, and has made His bounties flow to you in exceeding measure, (both) seen and unseen.

From the above verses, several relevant values regarding resources can be noted. Firstly, Allah (swt) is known as the ultimate Sustainer and Provider of all resources. Resources are endowed to men for the fulfillment of their daily needs and as a favor from Allah (swt). The treatment of such resources by humans should be akin to trustees who act in accordance to the teachings of Allah (swt). The notion of excluding some individuals as cultivated by Protestantism is in stark contrast and grave violation to this rule of the Creator.

Secondly, the adoption of any notions of scarcity of resources would be in violation to the Supremacy and powers of the Supreme Creator to create and provide for His Creations in just measures. Herein lies another difference between the Islamic and conventional model. Islam provides a different paradigm altogether to the notion of scarcity. In accordance to the Islamic model, any form of scarcity happens as a result of human misappropriation or inadequate endeavor (Alam, 1996). In Islam, there can be no scarcity at the macro level, only at the micro or local level. The Loving Creator Has given humanity all that is needed for flourishing societies *if* they share and refrain from opulent living and hoarding of wealth. The emergence of scarcity at the micro level is a test for humanity—to share as life on this earth is short and is intended to prepare for the hereafter. Hence, the Qur'an places strong emphasis on upholding social justice and rules against waste, accumulation and concentration of wealth and extravagance (Mirakhor, 2009).

Finally, Allah (swt) stresses the importance of remembering His rules for the attainment of success. Rule compliance is a measure of the strength of one's Islamicity, and adherence to it determines the behavior of individuals and collectivities (Mirakhor & Hamid, 2009). The attainment of the "best reward" promised by Allah (swt) as a result of rule compliance should supersede the temporary gains derived from cheating or engaging in other forms of corrupt practices. According to Mirakhor (2009), rule compliance helps humans to serve their own best self-interest as no one knows best what these are other than the Creator. An example from the Qur'an was given to further explain this—asserting that there are things that humans may perceive as best for them but are in fact harmful, and there are things that humans believe to be harmful for them but are in fact not harmful. This assertion is immediately followed by "Allah knows best and you do not" from Surah Al-Baqarah, Verse 216 of the Qur'an (Mirakhor, 2009).

The preceding paragraphs provide the foundation elements behind the Islamic model so that the perspective on institutions according to Islam may be formed. The definition of institutions suggested by North is revisited below:

> set of rules, compliance, procedures, and moral and ethical behavioral norms designed to encourage the behavior of individuals in the interests of maximizing the wealth or utility or principles.

Differences in the definition according to Islamic teachings are outlined as follows:

1. Rule sets differ—in conventional literature they are driven by man and their societal historical conditions. Theories on institutions are

thus based on human behaviors. The rules according to Islam are those that are handed down by Allah (swt) and ordained not as a means to constrain man's behavior, rather to assist them so that they may reach their full human potential.

2. The pursuit of wealth does not signify the endgame; it is the pursuit of success in the Hereafter that carries more meaning for Muslims as the important and ultimate goal.
3. Trust in Allah for outcomes and provisions erases any apprehension. One's concern and responsibility in this plane of existence is to uphold righteousness according to the rules of Allah (swt). The unseen or future gains are left in their entirety to the will of Allah (swt).

The arguments put forth so far establishes that the mental make-up of a Muslim man along with the life choices that he makes should be in accordance to the rules of Allah (swt). The unintended consequences mentioned by North as a result of the choices that actors make to the detriment of society should be eliminated in the Islamic model. Based on the following sayings of the Prophet (sawa), individuals in the Islamic system are charged with the responsibility of upholding social justice and welfare of the community along with its protection (Alam, 1996):

> A believer (in Islam) to a brother believer is a support, as each part of a building supports the other.[2]

> You will recognize the faithful by their mutual compassion, love and sympathy. They are like one body; if one of its parts is ill the whole body suffers from sleeplessness and fever.[3]

At this juncture, it is worthwhile to note that whilst relevant and complete guidance is given to equip Man (from the Qur'an and from the intellectual ability to appreciate and understand them), tremendous importance is also given to freedom. Islam places importance on the freedom of speech, freedom of action, freedom of association and freedom of religion and that is why Allah (swt) has assigned Muslims the duty to *enjoin the good and forbid*

2. Alam (1996), p. 27.
3. Alam (1996), p. 28.

evil (Alam, 1996). This message can be seen in Surah Al-Imran, Verse 104 of the Qur'an:

> And from among you there should be a party who invite to good and enjoin what is right and forbid the wrong, and these it is that shall be successful.

This duty is in a sense the capstone rule in Islam. It is the rule that overarches all other rules. This duty of enjoining good and forbidding evil cannot be performed without total freedom. Freedom is Allah's greatest gift to humans. If The Almighty, the all powerful, had wanted to create all humans perfect He could have done so. But then without the freedom to choose, life on this plane of existence would have lost all meaning. Life on this earth is a test for humans to develop for the Hereafter. And Allah has provided humans with the perfect guideline, namely the Holy Qur'an with detailed rules and their interpretation by the Prophet (sawa), for achieving perfection. Thus, any unintended consequences that arise are the effects of swaying away from the rules of Allah (swt) and not from the inadequacies of the rules themselves. This is a stark contrast to the conventional system.

Islamic Institutional Economic Framework

The necessary condition for the ideal economy within the Meta-framework is *iman*, being rule compliant. Men are guided to establish *taqwa*, the activated inner torch of continuous awareness and consciousness towards the Supreme Creator in order to be fully rule compliant. Members of the community comply with rules that are generally specified for all and specific rules that have been prescribed to in circumstances relating to economic behavior. These include the rule of enjoining good and forbidding evil, establishing social justice, establishing mutual cooperation and care towards each other and avoiding harm to others. The Islamic institutional framework for the ideal economy is made up of a collection of institutions of which its rules and conduct are prescribed by the Metaframework and operationalized through the Archetypal Model to govern matters pertaining to allocation of resources, production and exchange of goods and services and distribution and redistribution of income and wealth. The primary objective of these institutions is to achieve social justice.

The exemplar of successful implementation of the Metaframework relates to the period when Prophet Muhammad (sawa) operationalized its rules through an Archetypal Model for the society in Medina (Mirakhor, 2009). According to Mirakhor, the Archetypal Model defining the economic institutional framework prescribed in Islam includes sanctity of contract, property rights, trust, rules of behavior in governance, existence of markets, rules regarding production and distribution of resources, income and wealth; rules governing the behavior of market participants and rules regarding post-market distribution.

The author's explanation regarding these elements (supplemented by similar discussion points by Alam) together with the supporting Qur'anic verses and Sunnah of the Prophet (sawa) is summarized below:

1. Sanctity of contracts—Contracts are ordained in all relations pertaining to social, political and economic matters. The Qur'an urges believers to fulfill all their contractual obligations. This message is clear in the following verse:

   > Surah An-Nahl, Verses 91–92:
   >
   > Fulfill the covenant of Allah when you have pledged to do so, and do not break your oaths after asserting them, and after you have made Allah your guarantor (that you will perform the obligation entered into). Allah knows all that you do. And do not be like a woman who unravels her yarn to thin filaments after its strands have been firmly spun, you use your oaths as means of deceit to snatch at advantages over one another by violating agreements that have been sealed simply because some of you may be more powerful or numerous than others. Indeed, in this Allah tests you. And He will undoubtedly explain to you on the Day of Judgment what you have been disagreeing about.

   Allah (swt) commands it in the Qur'an in the same way that Allah (swt) commands believers to fulfill their covenant with Him. Thus, fulfilling contractual obligations consequently becomes a reflection of faithfulness to Allah (swt). From the birth of Islam, the body of rules based on the Qur'an and the Sunnah of Prophet (sawa) is made up from the theory of contracts. Islam forcefully anchors all of its socio-political-economic relations on contracts and its application and actualization forms the very basis of the Divine Law (Mirakhor, 2009). It represents the foundation of the covenant between the Creator and man, fulfilling contractual terms that represent the affirmation that man recognizes the Creator as his Cherisher

Lord. This proposition is also indicative of the links between humans, their Creator and to one another.

Mirakhor and Hamid (2009) explain that the body of rules covering all contracts has established the principle that any agreement not specifically prohibited by the Law was valid and binding on parties. These contracts are to be enforced by the courts who shall treat the parties to a contract as complete equals. It is an important rule of social interaction, this command of faithfulness to the terms of contracts. Every contract entered into by believers must be supported by the intention to remain loyal to performing the specified obligations. It is seen as a clear tool to protect and define property rights.

> Problems of credit commitment are stated to arise because parties involved in an exchange are not able to commit themselves or they do not have credible belief that others can commit themselves to carry out contractual obligations. The proper enforcement will result in substantial long-term benefits, such as better cooperation, coordination, trust and sustained long-term projects that can be supported at lower transaction costs.

2. Property rights: All properties and resources belong to Allah (swt). They are held as a sacred trust and amongst the rules ordained in its management include protection against exploitation through unjust and wrongful dealings. A claim to property is a result of one's labor, gift, inheritance and exchange. Natural resources are to be placed in the trust of either the state to be used by all, or in the hands of society at large. Producers are to ensure that there are no external adverse impacts of private economic behavior on society. The author provides the following Prophetic rule that was promulgated to ensure this:

> No harm. No injury. (lā darar wa lā dirār)

Instantaneous property rights such as those commonly in the form of usury or interest are strictly prohibited and property gained through unjust dealings make them unlawful for consumption. According to Mirakhor (2009) there are only two ways in which individuals can claim legitimate claims over property rights and that is through the combination of their own labor and other resources or through contracts of exchange, grants or inheritance from others who have gained property rights on assets obtained through their own labor.

Work therefore, is the ultimate premise and duty to gain rights over property.

The Qur'an places great importance on pushing members of society to work, as work also reflects worship of the Creator. The complementary rule from the Qur'an in relation to work is how instantaneous gains claimed over property rights are forbidden without commensurate work. An exception is transfers of gifts. The claim of property right over resources is given to all individuals, even those who are less able in the society. Allah (swt) ordains sharing and imposes limits over the disposing and treatment of property. Property owners have an obligation to the Creator not to waste, squander, destroy or use opulently the resources given to them. In this respect, resources are seen as sacred where none has the right to force its appropriation or expropriation.

3. Trust—The Qur'an and the Prophet (sawa) place great emphasis on trust. Trustworthiness demonstrates true belief and the betrayal of it, as a sign of disbelief. Irrespective of whether parties enter into an agreement with friend or foe, the trust that is undertaken is not to be violated. Trust is interconnected with contracts in many ways as without it, contracts become difficult to negotiate, conclude and enforce. High degree of trust reduces transaction costs and thus lubricates the wheels of business and commerce. The following verse from the Qur'an delivers the message on the importance of trust:

> Surah An-Nisa, Verse 105, 107:
>
> Truly, We have revealed to you (O Muhammad) the Qur'an with the truth, so that you may judge among men by which Allah has shown you, through his revelation. And you should not plead for those who are false in their trusts. … Nor shall you argue on behalf of those who deceive themselves; for surely Allah does not love the treacherous and the sinful.

The messages from the Qur'an are distinct and clear in showing that trustworthiness and remaining faithful to all contractual obligations and promises are absolute and binding. Social solidarity, mutual cooperation and brotherhood break when there is no trust between one another. Trustworthiness and keeping to obligations of contracts between man mirror how one maintains faithfulness to the Covenants with Allah (swt), including everything that the Creator has placed in trust on the believer. These include the resources that He has blessed upon the believer and the trust entails managing all resources according to His rules and guidance.

4. Rules of behavior in governance: The fundamental and principal objective of the rules of Allah (swt) is to achieve justice on this plane of existence. The function of rules is to reduce uncertainty for individuals. It specifies the expected conduct in a man's daily life, and in his interpersonal dealings. It is meant to control the inherent internal reactions that men would have from their emotions, passions, whims and impulses. These rules are specified to impose restrictions and ensure that appropriate conduct is maintained for society's social order. This will result in reconciliation of differences, coordination in action, cooperation and prevention of conflicts. The Qur'anic verse below distinctly shows that unity and order are fundamental for social cohesion:

> Surah Al-Imran, Verse 103:
>
> And cling fast one and all to the bond of Allah (the Religion of Islam), and do not draw apart from one another. And remember the favors He has bestowed upon you: how He united your hearts when you were enemies (during the Age of Ignorance), so that you come as brothers (in Islam) through His grace; and how He delivered you from the abyss of Fire when you were on the very brink of it (because of your unbelief during the Age of Ignorance) and He saved you from it. Thus Allah makes clear to you His revelations, so that you may be rightly guided.

Social unity is stated to be crucial and central as all rules of behavior ordained by Allah (swt) guide man to ensure its preservation. On the other hand, all rules that are prohibited are to curb any behavior that would jeopardize this unity. All creations are equal and man is of one kind. Allah (swt) explains in the following verses the fundamental message that the creation of man is from one Self, and their return is as one Self:

> Surah Al-Imran, Verse 28:
>
> Neither your creation (was) nor your resurrection will be other than as one united self.
>
> Surah Al-Ankaboot, Verse 29:
>
> As He brought you into being, so will you return.

This unity mirrors the reflection of the unity of the Creator. This, in turn, demonstrates the *wālayah* (loving care) of humans for one another. The progression of belief would entail full surrender of the will of oneself to the Will of Allah (swt) and full compliance to the will He has prescribed. *Iman* (belief) motivates love for humanity, which, in turn, empowers rule compliance. The conduct of a Muslim in governance is to command the good and forbid transgressions, ultimately to achieve and maintain social justice. The management of economic transactions and sharing of economic risks are all motivated by *wālayah* for one another to ensure that these interactions are done in the best possible manner. Alam (1996) explains this unity in the form of brotherhood and views members of mankind as one single family created by Allah (swt) from one couple, Adam and Eve. This brotherhood rejects any geographical boundaries or notions of elitist or preferential treatment of one tribe or race over another. Difference between individuals is in the level of righteousness. The following verse from the Qur'an explains this message:

> Surah Al-Hujraat, Verse 13:
>
> O mankind! We created you from a single (pair) of a male and female, and made you into nations and tribes, that you may know each other. Verily the most honoured of you in the sight of Allah is (he who is) the most righteous of you.

The notion of brotherhood in economic terms, as explained by Alam, implies that everyone gets his due for his or her contribution to society. The income due to everyone is obtained in the most just manner and not by depriving others of their share through exploitations. It refers to providing equal opportunities, and sharing the risks and rewards together as one human brotherhood. On the other hand, a gross difference in income and inequality destroys, rather than fosters, the feelings of brotherhood. Every person is thus free to search livelihood from anywhere and to adopt any lawful means for acquiring wealth. Any inequalities that do persist are considered as natural disparities; and these are duly accounted for in the distribution and redistribution mechanism in Islam. Inequality arising from artificial disparities refers to that which is created by man, and its existence reflects the break in the unity of humankind.

5. Rules governing market—Markets are the only authorized place of trade with the state made responsible for their construction, supervision, rule enforcement and maintenance. There should be no barriers to entry and exit, information regarding price, quality and quantity is to be made transparent and contracts should fully specify the property rights of all participants. State and governing bodies are responsible to guarantee contract enforcement and to ensure that hoarding of commodities is prohibited. The right to terminate a deal is given importance to ensure that market participants are protected against faulty information. During the time of the Prophet (sawa), a market supervisor governed the market to ensure there was no cheating, fraud, withholding of information and other malpractices that would jeopardize the natural flow of the market. Natural competition drove the price mechanism of the market, and a *muhtasib* (market supervisor) was placed at every market to ensure fair practices and to hold participants accountable for any misdealing.
6. Distribution—The fulfillment of rights and entitlements of other members of society are assured through payment of levies and voluntary contributions, such as *khums* (on income), *zakah* (on wealth), *infaq* (spending in the way of Allah [swt]), *qard hassan* (benevolent loan), *sadaqat* (redemption from others who were made less fortunate) and *waqf* (designated assets whose underlying income flows are used to support public infrastructures). *Zakah* is also imposed to provide an educative role to the individual so that he or she may be cognizant of his or her role in society. It is levied on net worth and not income, thus it is effective in transferring part of the wealth from those who have wealth to those who do not. The problem plaguing contemporary economics is finding an optimum distribution as there is no satisfactory theory that can link this with marginal productivity to effect a functional income distribution (Alam, 1996). The beauty of the rules of Allah (swt) is that they provide the best solution to achieve optimal distribution. They enjoin the wealthier groups of society who enjoy surpluses in income to make the less able groups co-sharers in their wealth through the forms of obligatory and voluntary means of distribution. This educative role directly affects their mindset and triggers the action to work hard so that individuals may obtain enough income, for their direct needs, the needs of their families and for the allocation of *zakah* and other voluntary contributions.

The result of this transfer creates an upward shift in the aggregate demand function because the marginal propensity to consume of those who receive the transfer payments is comparatively higher (Alam, 1996). The distributive role of *zakah* is significant as it allocates funds to those in need of essential goods and services and diverts the waste that would have otherwise been spent on luxurious items and on items beyond a person's needs. The income multiplier effect it provides takes the form of motivating the investor to deplete idle stock of capital into productive use. It discourages hoarding and accumulation of idle wealth. Resources are also channeled back into economic activities as demand for consumption increases.

Another powerful tool in distribution refers to *qard hassan*. It was practised by the Prophet (sawa) to transfer funds to a person in need so that the person's financial hardship may be eased. The loan is free from interest and postponement of loan recovery is encouraged. The latter is meant to ensure that the debtor truly regains better financial standing. The demonstration of trust, compassion and empathy from the provider of the loan are meant to strengthen social solidarity and mutual cooperation. The Prophet (sawa) is reported to have said the following:

> May Allah have mercy and blessings upon a man who is kind-hearted and generous in his dealings and behaves gently in demanding his loans.[4]

These words thus explain that delayed recovery from the debtor is encouraged but remission of debt is even better. During the time of the Prophet (sawa) the loan could also take the form of capital assets given to a person in need for his or her benefit during the agreed fixed period. These assets include money, animals, cattle for milk, lands, fruit-producing trees and accommodation in houses. In the spirit to remove a member of the society from destitution, the significance of *qard hassan* is that it serves as an important distribution mechanism to increase the purchasing power of individuals and to narrow the gap between those who have surpluses of income to those who do not.

> Spending in the way of Allah (swt) is further explained in the Qur'anic verse below:

4. Alam (1996), p. 179.

Surah Al-Baqarah, Verse 215:

> they ask you what they shall spend. Say: That which you spend for good (must go) to parents and near kindred and orphans and the needy and the wayfarer. And whatsoever good you do, Lo! Allah is aware of it.

This expenditure is a voluntary contribution that is given additional to the *zakah* and other voluntary contributions. It can also be referred to as *sadaqat* and the sincere contribution is reflective of one's sincerity and the strength of his belief in Islam. Allah (swt) elucidates the following groups of people as the recipient of these contributions in the following verse from the Qur'an:

Surah Al-Baqarah, Verse 273:

> (Sadaqat) are for the poor who are in dire straits in the way of Allah, who are unable to use the land (either for traveling in search of trade or to farm the earth), The ignorant (unthinking) think them wealthy because of their restraint and modesty (in revealing their poverty). You will know them by their mark: they do not beg from the people, covering (their poverty), And whatsoever good things you spend, lo! Allah knows it.

In another verse (see below) Allah (swt) reminds man to be sincere and kind towards the needy so that they may be treated with dignity. The verse also serves as a reminder that the reward of giving provides a descent of blessings that increases one's wealth instead of decreasing it. Seeking the pleasure of Allah (swt) should be the primary concern and giving should not be done to obtain popularity or for show, and it is encouraged to be done in secret only for the knowledge of Allah (swt).

Surah Al-Baqarah, Verses 262–265:

> Those who spend their wealth for the cause of Allah and do not accompany (the expenditure) with reproach and (hurting the feelings of the recipient with) annoyance and ill-treatment, their reward is with their Cherisher Lord and no fear and grief shall come upon them. A kind word with forgiveness is better than a sadaqah given with reproach and ill treatment (of the recipient), like the one who spends his wealth only to be seen (to show off) by other humans while he does not (really) believe in Allah and the Last Day. His giving is like that of dust settling on a solid rock, when the rainstorm hits it (the rock) and washes it (the dust) away leaving the rock cleansed

and bare. These people have no control over what they have gained. And, Allah does not guide the disbelieving folk: (On the other hand), the likeness of those who spend their wealth only to seek the pleasure of Allah and to strengthen their self (to develop the self), is as the likeness of a highly placed garden; (when) the rainstorm hits it (the garden) brings forth its fruit twofold. And if it is not a rainstorm that hits it, then it will be a shower (that pours on it). And, Allah is All-seeing of what you do:

Another profound benefit of spending in the way of Allah (swt) is that it is able to purify, or cleanse past transgressions so that the man's slate on this plane of existence can be purified. This message is mentioned in the verse below:

Surah Al-Baqarah, Verse 271:

If you make the payment of sadaqat (redeeming the rights of others in one's wealth) known to others, it is well, but if you hide it (give in secret without revealing to others) while giving to the poor, it will be better for you, and will atone some of your transgressions. And, Allah knows well what you do

Alam (1996) provides the following message from the Prophet (sawa). It portrays with impact that turning one's back to those in need is likened to turning against Allah (swt).

The Prophet (sawa) states that Allah (swt) will say the following on the day of resurrection:

"O Son of Adam I was ill and you did not succor Me," Man will explain: "O! Lord! How could I have succored Thee, the Lord of all the worlds and Allah will reply: Did you not know that such and such of My servant was ill, and you did not succor him? Did you not know that if you had done so, you would indeed have found Me with him." "O Son of Adam! I asked you for food but you did not feed me," Man will therefore say: "O Lord, how could I have fed Thee, the Lord of all the worlds?" Whereupon Allah will say: "Did you not know that such and such of My servants asked you for food, and you did not feed him? Did you not know that if you had done so, you would indeed have found it (again) with Me? O Son of Adam! I asked you for a drink but you did not give me a drink." Man will say thereupon: "How could I have given Thee, the Lord of all the worlds, to drink?" But Allah will reply "such and such of My servant asked you for a drink, but you did not give it to him. Did you not know that if you had given him to drink, you would have found it (again) with Me."[5]

5. Alam (1996), p. 163.

The distribution and redistribution mechanisms in Islam work on the objectives of establishing a just economic system, enabling everyone to get what is rightfully due to him or her and preventing the concentration of wealth. Payments of *sadaqat* reflect the strength of the believer's sincerity to Allah (swt). Extreme care is to be taken to preserve the dignity of the recipient of *sadaqat* contributions, and even encourage that it is made in the most discreet manner. Showing any ill treatment or displeasure towards the recipient is strictly forbidden as the contribution should be seen as a rightful claim that the least fortunate have over those that have been blessed with more wealth. Some have been privileged with greater physical and mental abilities while the conditions of others are not as fortunate. Wealth, should not just circulate within the rich.

The Prophet (sawa) warned about the evil consequence of inequitable distribution of income and wealth. He is reported to have said the following:

> He who hoards is the worst criminal.[6]

> He who hoards food grains for forty days to raise prices, is in fact dissatisfied with Allah and Allah is also dissatisfied with him.[7]

The provisions given by the Creator are sufficient for everyone to draw their sustenance. Due to differences in capability, Islam makes it obligatory on society and its rulers to look after the needy and to fulfill their needs. The remaining wealth is to be distributed to the relevant beneficiaries in accordance with the rules of inheritance. The inheritance rule in Islam is seen as an effective means for distribution of wealth and property and to avoid the concentration of wealth beyond a person's justified needs. It rejects the unjust principles of primogeniture or male joint family and bequest through wills to relationships that are favored. The disposing of one's wealth correlates the interest of the family with his societal rights of being a member of the society to chart a wider path of circulation.

7. Production—Humans are urged to work hard in combining their physical and mental labor with the resources provided by the Creator

6. Alam (1996), p. 73.
7. Alam (1996), p. 74.

to produce goods and services and wealth for the benefit of themselves and mankind. Followers are also encouraged to work for themselves and to develop their own life of productive activity rather than work for others. The individual's life should be arranged so that he provides responsible creation for the society and does not spend time in unproductive and inactive occupation. Entrepreneurs are subjected to the rules of economic behavior of not cheating and engaging in waste. Laborers are to be given dignified respect and treatment.

8. Rules on income and wealth: The Islamic vision for income and wealth prohibits wealth concentration to circulate at the expense of the wider society, where there are people in need and there is gross inequality in income. Wealth and resources are seen as blessings from Allah (swt), thus one's sincerity in one's belief would make them acquire and spend it according to how Allah (swt) has commanded. Although man engages himself in the process of production, it is Allah (swt) who can cause the endeavor to reap rewards and become fruitful. This message is given in the following of Surah Al-Waqia, Verses 63–64 from the Qur'an.

> Have you considered that you till? Is it you yourselves who make it grow, or is it We Who make it grow.

In order to allow for a healthy circulation of wealth, the financial system should be based on risk and return sharing. Interest-based transactions on the other hand would shift the risks to the borrowers. Islam proposes *al-bay*—mutual exchange where a set of property rights is exchanged for another, allowing for both parties to justly share the risks of the transaction. The nature of property rights is what separates between interest-based and *al-bay* exchanges. The exchange and trade of commodities or assets are said to be the foundation of economic activities in Islam. Exchange requires freedom of parties to enter into a contract. Consequently this leads to the freedom to produce and allows for clear and protected property rights to permit production. Sale of production, in turn, requires markets to enable buyers and sellers to conduct their mutual exchange. Financial resources thus do not go to the hands of those that are already rich but would instead reach entrepreneurs.

Another feature of the risk-sharing return-sharing mechanism is that the rate of return for providing financing is determined *ex-post* and by the rate of return to the *real* economic activity. An interest-based system employs the reverse as debt contracts finance production. The risk-return sharing format removes interest payments from the production phase and places them in the post-production phase, including the distribution phase, which takes place after sales. This, in turn, triggers a direct price to quantity effect. The rules prescribed by Allah (swt) reduce risk and uncertainty including the risk that arises from income that causes difficulty in people's livelihood. The trade and exchange in commodities and assets are designed to promote risk sharing and to create the potential for enhanced social solidarity. Risk-reward sharing is one of the most fundamental conditions for all economic relations such as governing resource allocation, property rights, production, exchange, distribution, redistribution, financial transaction and market behavior.

As an aside, the recommended financial system in Islam has the additional benefit of promoting financial stability. Under this financial system, banks provide two financial services—safekeeping of deposits and mutual funds/investment banking. Thus, in taking on safekeeping deposits, banks assume no risk, as these funds are not lent (banks do not create money). In their role as providers of mutual funds and investment banking services, banks invest a client's deposits directly on a pass-through basis without risking any of their own capital. The bank only risks its own capital if it also invests in a project. In essence, banks are not leveraged, risking multiples of their capital in loans and investments. In other words, this is similar to 100% reserve banking.

The arrangement of the system is complete and guarantees justice and fairness in the sharing of risk and reward. Rent-seeking behavior is thus a stark violation of the risk-reward sharing mechanism. Once income and wealth have been obtained, the program for redistribution of income and wealth is complete to guarantee a standard of living for every individual that would be commensurate with his dignity on earth. However, if there are members of society in need, then everyone who had more than his or her basic needs is obliged to engage in redistribution to promote the welfare of society. There is no room for hoarding or excessive spending at the expense of other members of society. This point is elucidated in the following verse of the Qur'an;[8]

8. Other related verses are listed in Appendix 3b.

Surah Al-Maarij, Verses 24–25:

> In their wealth there is a known right for those who ask for it and those who have need for it.

The prescribed method for expenditures avoids miserliness and extravagance by consciously exercising moderation in spending.

9. Rules governing market behavior: Markets are seen as instruments whose structures are designed to possess the required characteristics to promote rule compliance. Prices should be determined by the competition among suppliers. A person of authority who will hold market participants accountable for any misconduct or unfair practices would supervise markets. The behavior of market participants is subject to the rules prescribed by Allah (swt). They are to be internalized before entering the market. Such rules prohibit *israf* (overspending), *itlaf* (waste), and *itraf* (ostentatious and opulent spending), *lā darar wa lā dirār* (no harm or injury to anyone), being faithful to contracts, as well as practicing trustworthiness. Market rules specific to exchange such as no fraud, cheating, shortchanging of weights and measures, interference with the natural form of demand and supply, hoarding of commodities to raise prices and restriction on information also apply. Markets that operate in full rule compliance will result in just prices for goods, services and reward for factors of production.
10. Rules governing redistribution: Government intervention is required to correct any unjust pattern of distribution. The state is expected to take the role of an administrator, supervisor and protector of the society. In contrast to that of the capitalist and Marxist systems, Islam's own unique system promotes justice as an important outcome. The primary objectives include enabling everyone to get what is rightfully due to him, and preventing concentration of wealth. According to Alam (1996), in a truly Islamic society there is little likelihood of any divergence between social and individual good. However, should any such instances arise, the state should intervene, educate public opinion to bridge any divergence from Islamic teachings, guide and regulate production and distribution to ensure that the allocation of resources conforms to the goals of Islam.

To summarize, the above definitions for institutions and institutional framework in Islam are clearly and distinctly different from the definitions in conventional systems. The Islamic institutional framework is built on morality and justice to establish thriving societies where everyone has an equal opportunity. The system is based on an intricate web of rules that are promulgated to ensure that humans individually and collectively realize their full potential and to prepare them for a more meaningful life in the Hereafter. The rules are also designed to ensure that men do not produce collective ills and cause disorder on earth.

The Qur'an and Sunnah of the Prophet (sawa) are rich and full of messages and guidance, which aim to ensure that the livelihood of all members of the community is protected. Those that are mentioned above do not represent the majority of the guidance. The many different tools that man can use for distribution of wealth are put in place to ensure that everyone has the means to obtain sustenance and live a dignified life. There is extensive attention, affection and consideration given for those in need as seen from the verses of the divine Qur'an.

As promised by Allah (swt), the full operationalization of His rules would result in rewards both in this world and the Hereafter. As it is not man-made, it becomes free from the flaws that man face. The reward of sustained provision beyond what one could ever measure would result in economic prosperity leading to just and united societies.

Do Institutions Support Sustained Economic Growth?

The different belief systems underlying the notion of institutions as suggested by Douglass North shape the different economic performances and results across different societies and countries. The Islamic system goes much further than North to recommend rules that promote morality, trust, compassion and coordination. The Islamic system prohibits vast accumulation of wealth and recommends sharing to build social solidarity. While there is more resemblance to the system perceived by Adam Smith than was premised on his less referenced book, *The Theory of Moral Sentiments*, even his vision did not go far enough to create a harmonious, just and thriving economy that served all humanity. As an extension to this argument, from the Islamic perspective it is the adherence to Islamic teachings and practices that shapes and determines economic performance and its lack of adherence that causes the differences, if any. Thus, according to the principles of Islamic economics, institutions will indeed support economic growth, and will go as far as to guarantee it if the institutional framework is one that is completely aligned to the rules of the Creator. Sub-par economic performance and economic failure are results of rule non-compliance.

Why then are such discrepancies noted in the studied countries with the resource curse, and slow economic growth and institutional weaknesses raise cause for concerns? If the Islamic vision guarantees substantial benefits, why do the oil-producing countries not exhibit it? Why do poverty and income inequality still persist in the countries—albeit the rich and comprehensive guidance from the Qur'an and Sunnah? Are greed and corrupt practices hidden behind their professed Islamic administrations? Why is there such oppression and injustice? Why is there no respect for the rule of law? The next section will attempt to answer these questions by reviewing the prevailing institutional state of these countries.

Institutional Condition of Muslim Countries

The discussion put forward thus far has established the following; that if the oil-producing countries had effective institutions when they began to produce oil or if they had managed to adopt effective institutions even after they discovered oil—something that was and is harder to do after the discovery of significant oil deposits under the control of the government—oil would in all likelihood have been a blessing. And if they had been compliant with the Divine rules of conduct in the management of resources, then they would have been able to create just and thriving societies. The Islamic economic institutional framework—with higher levels of trust, coordination, cooperation and sharing—would demonstrate superior performance in their economies, and improved well-being of their societies. Are the countries in conformance with Islamic rules? The Islamicity Index ("the index") has been developed as an instrument to answer such questions.

Developed by Askari et al. in 2006, it has been updated by Askari and Mohammadkhan (2015) and more recently enhanced, improved and annually updated by Askari, Mohammadkhan and Mydin (2016), the index is the first of its kind to provide scores and ranking for a country's adherence to Islamic teachings. The authors suggest that countries which have truly adopted the Islamic vision would reflect the following characteristics in their economic system; greater degree of justice in all aspects of economic management, higher moral standard, honesty and trust exhibited in the marketplace and in all economic transactions, poverty eradication, a more even distribution of wealth and income, no hoarding of wealth, less opulence in consumption, no exploitive speculation, risk sharing contracts as opposed to debt contracts, better social infrastructure and provision of social services, better treatment of workers, higher education expenditures relative to GDP, higher savings and investment rates, a higher degree of environmental preservation and vigilantly supervised markets. They further explain that successful implementation of Islamic rules would also be reflected in higher qualitative and quantitative economic growth.

The index is made up of important dimensions that are based on the foundational Islamic teachings from Qur'an and the practices of the Prophet. These dimensions encompass the areas of economic, legal and governance, human and political rights, and international relations which are used to assess how well Muslim countries adhere to the Islamic teachings that they claim to uphold. It reports the results for 153 countries across all four of the dimensions and a final aggregate ranking is provided in an overall Islamicity index score.

Overall, the report found that out of 153 countries not a single Muslim majority country made the top 25 ranking. The self-declared Islamic countries did not fare very well, indicating their lack of adherence to Islamic principles listed by Askari and Mohammadkhan (2015). The results indicate that Western countries, especially a handful of northern European countries, Canada, Australia and New Zealand, perform the best in the indices. If we turn to the results in more detail, we do not find any representation of Muslim countries in the top ten across any of indices (Tables 4.1–4.5). New Zealand, Canada, Australia and the Northern European countries dominate the top 10 rankings.

In the Economic Islamicity Index (Table 4.1), Denmark is the best performing country with a score of 9.002. According to Askari and Mohammadkhan (2015), the top performing countries effectively demonstrate sound institutions and freedom and opportunities for the individual to develop. These countries also have extensive social programs and have excellent state-funded education systems, an essential element for individual and societal economic development.

Table 4.1: Economic Islamicity Index (Top 10 Countries).

Country	Rank	Score
Denmark	1	9.002
Switzerland	2	8.933
Netherlands	3	8.886
Austria	4	8.656
Finland	5	8.626
Sweden	6	8.625
Germany	7	8.506
Canada	8	8.416
New Zealand	9	8.384
Norway	10	8.340

Source: Askari and Mohammadkhan (2015).

The Legal and Governance Islamicity Index (Table 4.2) in essence covers governance and judicial assessments and looks at the state of legal integrity, perceptions about government and government governance. Again, the index puts Western countries at the top 10, with Switzerland as the best performing country at a score of 9.559. Singapore shows a 9.009 score, a strong performance at number 5.

Table 4.2: Legal and Governance Islamicity Index (Top 10 Countries).

Country	Rank	Score
Switzerland	1	9.559
Luxembourg	2	9.493
New Zealand	3	9.267
Netherlands	4	9.031
Singapore	5	9.009
Denmark	6	8.943
Finland	7	8.935
Norway	8	8.775
Sweden	9	8.731
Germany	10	8.723

Source: Askari and Mohammadkhan (2015).

In the case of the Human and Political Rights Islamicity Index (Table 4.3), the top rankings, are predominantly made up of the same Western counties as the previous two indices. The results are not a surprise, as these countries have long been known as proponents of human development and civil and political rights.

Table 4.3: Human and Political Rights Islamicity Index (Top 10 Countries).

Country	Rank	Score
Norway	1	9.850
Sweden	2	9.699
Denmark	3	9.529
Finland	4	9.438
Netherlands	5	9.340
Iceland	6	9.314
Canada	7	9.222
Australia	8	9.176
New Zealand	9	9.157
Switzerland	10	9.150

Source: Askari and Mohammadkhan (2015).

In the International Relations Islamicity Index (Table 4.4), the high performing countries differ from the three previous indices. Iceland tops the ranklngs with a score of 8.660. Muslim countries still are not included in any of the top-ranking positions.

Table 4.4: International Relations Islamicity Index (Top 10 Countries).

Country	Rank	Score
Iceland	1	8.660
Malta	2	8.627
Ireland	3	8.497
Panama	4	8.235
Netherlands	5	8.105
Luxembourg	6	7.876
Costa Rica	6	7.876
Sweden	8	7.810
Belgium	9	7.745
Czech Republic	10	7.582

Source: Askari and Mohammadkhan (2015).

In the overall Islamicity Index (Table 4.5) and as to be expected, Western countries dominate the top positions with Netherlands leading the set with a score of 8.909. Thus, it can be concluded that the top 10 ranking of developed non-Muslim Western countries reflects Islamic precepts much better than Muslim countries do.

Table 4.5: Overall Islamicity Index (Top 10 Countries).

Country	Rank	Score
Netherlands	1	8.909
Sweden	2	8.865
Switzerland	3	8.778
New Zealand	4	8.726
Denmark	5	8.651
Finland	6	8.573
Norway	6	8.495
Luxembourg	8	8.478
Australia	9	8.443
Canada	10	8.437

Source: Askari and Mohammadkhan (2015).

We now turn to the results of the sample OIC countries in our study (Tables 4.6–4.10). The Economic Islamicity Index is listed in Table 4.6 where the best performing country Malaysia is at number 38 with a score of 6.642. We see this score as significantly lower than the top-ranking score held by Denmark at 9.002. Other countries in the studies did not fare well as Middle Eastern countries in the group ranked in the 40s and other countries such as Indonesia were on the 90th rank and lower. Cameroon and Nigeria dominated the bottom rankings in the group. These indices indicate unsatisfactory achievement in economic justice, broad-based prosperity and job creation. Their policies, economic governance and achievements fall short and are significantly behind developed Western countries mentioned previously. The results are, of course, not surprising as we have previously established that institutional failure in Muslim countries have impeded any efforts in achieving sound economic and distributive justice.

Table 4.6: Economic Islamicity Index (Countries Included in This Study).

Country	Rank	Score
Malaysia	38	6.642
Bahrain	41	6.415
Saudi Arabia	47	6.000
Oman	44	6.240
Indonesia	90	4.195
Algeria	93	4.171
Cote D'Ivoire	103	3.893
Egypt	112	3.612
Nigeria	124	3.362
Cameroon	129	3.193

Source: Askari and Mohammadkhan (2015).

Again, the Legal and Governance Islamicity Index (Table 4.7) affords the countries in the study a low ranking. Although Malaysia ranked higher in this index, claiming the 24th rank at a score of 7.456, other countries still fell in the bottom ranking. Lapses in the Nigeria's legal and governance framework are demonstrated by its 147th rank and a score of 1.757, an abysmal performance especially if we compare with the highest-ranking score claimed by Switzerland the 1st in the ranking at a score of 9.559 as we saw earlier.

Table 4.7: Legal and Governance Islamicity Index (Countries Included in This Study).

Country	Rank	Score
Malaysia	24	7.456
Bahrain	52	5.752
Saudi Arabia	40	6.424
Oman	48	6.070
Indonesia	82	4.520
Algeria	107	3.655
Egypt	123	3.113
Cameroon	139	2.328
Cote D'Ivoire	146	1.895
Nigeria	147	1.757

Source: Askari and Mohammadkhan (2015).

The Human and Political Islamicity Index (Table 4.8) saw the highest-performing country Malaysia fall to its lowest ranking thus far at the 76th rank and a score of 4.948.

We can see from the previous indices that the last four positions are mainly made up of an interchangeable placing between Egypt, Cameroon, Cote D'Ivoire or Nigeria. In the case of this index, Egypt holds the lowest rank at the 144th rank and a score of 1.863.

Table 4.8: Human and Political Islamicity Index (Countries Included in This Study).

Country	Rank	Score
Malaysia	76	4.948
Indonesia	79	4.830
Saudi Arabia	93	4.307
Oman	94	4.080
Algeria	99	3.791
Bahrain	102	3.712
Nigeria	132	2.458
Cameroon	139	2.209
Cote D'Ivoire	149	1.458
Egypt	144	1.863

Source: Askari and Mohammadkhan (2015).

The International Relations Islamicity Index (Table 4.9) however places Nigeria at the highest rank, at number 15 and a score of 7.222. This time, the last four positions are dominated by Saudi Arabia, Oman, Egypt and Algeria, a combination of countries that is different from the previous indices. However, because this index also incorporates the militarization prowess of the country, we are not surprised by the first two top positions being dominated by Nigeria and Cote D'Ivoire, countries that have constantly mobilized military apparatus as an immediate response to resolving conflicts.

The overall Islamicity rating of the countries in the study is outlined in Table 4.10. The ranking unsurprisingly shows that the countries did not demonstrate high adherence to the teachings and doctrines of Islam. More than half of the countries in study fell into the bottom 100 rating. Malaysia, as the highest ranking was only able to obtain the 43rd position and a score of 6.355.

Table 4.9: International Relations Islamicity Index (Countries Included in This Study).

Country	Rank	Score
Nigeria	15	7.222
Cote D'Ivoire	42	5.822
Malaysia	45	5.752
Indonesia	62	5.359
Cameroon	75	5.065
Bahrain	114	4.052
Saudi Arabia	118	3.954
Oman	132	3.300
Egypt	134	3.203
Algeria	144	2.288

Source: Askari and Mohammadkhan (2015).

Table 4.10: Overall Islamicity Index of the Countries (Countries Included in This Study).

Country	Rank	Score
Malaysia	43	6.355
Bahrain	63	5.260
Saudi Arabia	59	5.395
Oman	60	5.356
Indonesia	88	4.519
Algeria	113	3.580
Nigeria	129	3.163
Cameroon	134	2.956
Cote D'Ivoire	135	2.876
Egypt	137	2.726

Source: Askari, Mohammadkhan, and Mydin (2015).

In summary, despite having oil resources, the countries failed to show a high level of adherence to the principles of Islamic economics. This result supplements the finding in our study and provides supporting evidence that institutional weakness is indeed a cause for concern in Muslim countries. The current level of corruption perceptions of the countries is also analyzed. Based on the index by Transparency International, the corruption perceptions index measures the perceived levels of public sector corruption in 167 countries worldwide. Table 4.11 provides the ranking of the countries in study.

Table 4.11: Corruption Perceptions Index Ranking (Countries Included in This Study).

Country	Rank
Malaysia	48
Bahrain	50
Saudi Arabia	54
Oman	60
Indonesia	88
Algeria	88
Nigeria	88
Cameroon	107
Cote D'Ivoire	130
Egypt	136

The results paint a similar picture to the Islamicity indices. More than half of the countries fell in the bottom ranking of the index. Based on the report issued by the agency, a poor score is afforded when there is a sign of widespread bribery, lack of punishment for corruption and public institutions of the country do not respond to citizens' needs.

Askari and Rehman (2013) attribute the unsatisfactory performance of these countries to their bad governance, which they represent as corrupt, unrepresentative and oppressive. The rulers and their supporters are further said to indulge in self-enrichment with little incentive for developing and nourishing effective institutions. Governance is generally oppressive and is focused instead on excessive military expenditures so that the rulers may tighten their grip on power. Oil revenues are channeled to unproductive subsidies with the goal of buying domestic support. The authors further note that the effects of failed governance and ill-conceived policies have resulted in economic deprivation and injustice for the vast majority of citizens. They conclude, that the reason why Muslim countries have subpar performance is because they have not followed the basic teachings of Islam.

Clark (2016) reported that a study had unraveled a Saudi Arabian sovereign-wealth fund being prepared to take ownership of state-owned oil giant Saudi Arabian Oil Co. to be less transparent than a fund that was previously considered most opaque in the world, belonging to Equatorial Guinea. The author further reports efforts to obtain further information were futile when the query was made by a former very senior financial expert at the U.S. Treasury and Federal Reserve, Edwin Truman. The website of the Saudi Arabian fund mentioned that it has a portfolio of approximately 200 investments, of which around 20 are listed on the Saudi Stock Exchange; however, none of the investments are named. Truman compiles a ranking on the transparency of sovereign-wealth funds and found Norway, New Zealand, the United States, Azerbaijan and Canada disclosed the most information about its sovereign-wealth funds and Saudi Arabia, Equatorial Guinea, the United Arab Emirates, Libya, Algeria and Brunei disclosed the least.

Ibrahim, Mirakhor, and Ng (2015) offer further insights as to why the Muslim world is in a state of turmoil. The divergence in the current practices of Muslims from the vision discernible from the sacred sources of Islam is viewed as one of the root causes of the problem. Evidence of grave moral failures and unjust treatment of others and each other are forms of the divergence that are said to be rampant in the Muslim world. The conduct of Muslims is reported to be in the exact opposite of the beauty, serenity, mercy and compassion that define the spirit of Islam. The authors explain that there are two basic visions of Islam in competition in the Muslim world. One view is forced by a minority on the rest of the Muslims and is driven by and influenced by counterproductive behaviors arising from the greed for petrodollars. The other path is for Muslims to reclaim the vision of Islam that is clearly discernible from the Qur'an, which would lead them to peace, love, compassion and mercy. The authors further explain that the latter would entail an understanding and explaining of the Will and the Purposes of the Creator for His creation according to sources of Islam. In their view, reclaiming the vision in the latter is imperative, and to remain inactive in the face of grave moral failures is no longer an option for the Muslim majority.

Norway is an example of an oil-rich country that has done well in managing its resource abundance. According to Holden (2013), Norway does not fit in the group of countries plagued by the resource curse with natural resources having detrimental effects on economic performance and development. This is largely attributed to the effective institutional structure that is at the foundation of the country's economic practices and is aptly reflected in the management of oil resources. The author explains that Norwegian decision makers reflect the view that resources belong to the nation, and that the development should benefit the society as a whole, including future generations. A number of different policies were adopted to achieve this objective. Holden provides that the regulation and taxation system ensured oil revenues were exploited in a safe and efficient and effective way. Additionally, the government ensured a significant Norwegian participation in the petroleum activities by focusing on building-up expertise by Norwegian companies so that they are able to take part in the oil sector. Norway's political system is seen as one of the most stable in the world.

Holden explains that when oil was discovered in Norway, the country had been in a state of a stable democracy since it acquired independence in 1905. He further adds that the state bureaucracy and legal systems function well, with a minimal level of corruption. The government predominantly imposed heavy taxation on oil companies amounting to 78 percent tax on profits, and adopting this policy within the backdrop of well-functioning systems allowed the state to reap the bulk of the oil revenues. Although the taxation is heavy, the tax system is seen as stable and transparent, implying that oil companies have always viewed the Norwegian sector as attractive for business. Another crucial point suggested by Holden in resolving the resource curse concerns the link between oil revenues that accrue to the government, and the spending of these revenues. Diligent and transparent decisions ensure that investments are channeled to the best sectors for sustainable growth and are also reasons why Norway has been able to manage its resources well.

In adopting transparent, just, and fair distribution measures, the institutional framework adopted in Norway seems to reflect traces of Islamic values and principles. Indeed, Norway ranks amongst the top 10 of the Islamicity index. Resource abundance will remain a curse for as long as the countries in our study operate within the scaffolding of weak institutions and engage in practices that are contrary to the Divine teachings of Islam.

Conclusion

Whether in the conventional or Islamic context, there is telling evidence that highlights the need for change in the current institutional framework of many oil-producing countries. Institutional reforms are required to change the mental mindset, belief systems and incentive structures of economic and political agents and to minimize implementation of distortionary policies. But as we have said earlier, reforms that adversely affect the interests of entrenched oppressive rulers and other interest groups are difficult when access to vast natural resources is at stake. It would have been much better if oil had been discovered after these countries had established effective institutions! In the following chapter, we discuss our proposed recommendations for institutional reform.

References

Acemoglu, D., Johnson, S., & Robinson, J. a. (2000). The colonial origins of comparative development: An empirical investigation. *American Economic Review, 91*, 1369–1401. Retrieved from http://www.nber.org/papers/w7771

Acemoglu, D., Johnson, S., & Robinson, J. a. (2005). Chapter 6 institutions as a fundamental cause of long-run growth. *Handbook of Economic Growth, 1*(SUPPL. PART A), 385–472. https://doi.org/10.1016/S1574-0684(05)01006-3

Alam, D. M. M. (1996). *Perspective on Islamic economics* (1st ed.). New Delhi: Institute of Objective Studies.

Askari, H., & Mohammadkhan, H. (2015). *Islamicity indices: The seed for change*. New York, NY: Palgrave Macmillan.

Askari, H., Mohammadkhan, H., & Mydin, L. (2016). Islamicity Index. Retrieved from www.islamicity-index.org

Askari, H., & Rehman, S. S. (2013). A survey of the economic development of OIC countries. In A. Mirakhor & Z. Iqbal (Eds.), *Economic development and Islamic finance* (pp. 299–324). Washington, DC: World Bank.

Brunnschweiler, C. N. (2008). Cursing the blessings? Natural resource abundance, institutions, and economic growth. *World Development*, 36(3), 399–419. https://doi.org/10.1016/j.worlddev.2007.03.004

Bulte, E. H., Damania, R., & Deacon, R. T. (2005). Resource intensity, institutions, and development. *World Development*, 33(7), 1029–1044. https://doi.org/10.1016/j.worlddev.2005.04.004

Clark, S. (2016). Saudi wealth fund may be world's least transparent. *Wall Street Journal*, 2–3.

Costa, H. K. D. M., & Santos, E. M. Dos. (2013). Institutional analysis and the "resource curse" in developing countries. *Energy Policy*, 63, 788–795. https://doi.org/10.1016/j.enpol.2013.08.060

Glaeser, E. L., La Porta, R., Lopez-De-Silanes, F., & Shleifer, A. (2004). Do institutions cause growth? *Journal of Economic Growth*, 9, 271–303.

Holden, S. (2013). Avoiding the resource curse: The case Norway. *Energy Policy*, 63, 870–876. https://doi.org/10.1016/j.enpol.2013.09.010

Ibrahim, M., Mirakhor, A., & Ng, A. (2015). *Social capital and risk sharing: An Islamic finance paradigm*. New York, NY: Palgrave Macmillan.

Knack, S., & Keefer, P. (1995). Institutions and economic performance: Cross-country tests using alternative institutional indicators. *MPRA Paper World Bank, 23118*.

Mehlum, H., Moene, K., & Torvik, R. (2006). Institutions and the resource curse. *Economic Journal, 116*(508), 1–20. https://doi.org/10.1111/j.1468-0297.2006.01045.x

Mirakhor, A. (2009). Islamic economics and finance: An institutional perspective. *IIUM Journal of Economics and Management, 17*(1), 21.

Mirakhor, A., & Hamid, I. S. (2009). *Islam and development*. New York, NY: Lobal Scholarly Publication.

North, D. C. (1981). *Structure and change in economic history*. New York, NY: W. W. Norton & Company.

North, D. C. (1990). *Institutions, institutional change and economic performance*. Cambridge: Cambridge University Press.

North, D. C. (1993). Institutions and credible commitment. *Journal of Institutional and Theoretical Economics, 149*, 11–23. https://doi.org/10.2307/40751576

Oskenbayev, Y., Yilmaz, M., & Abdulla, K. (2013). Resource concentration, institutional quality and the natural resource curse. *Economic Systems, 37*(2), 254–270. https://doi.org/10.1016/j.ecosys.2012.11.001

Rodrik, D., Subramanian, A., & Trebbi, F. (2004). Institutions rule: The primacy of institutions over geography and integration in economic development. *Journal of Economic Growth*, 9(2), 131–165. https://doi.org/10.1023/B:JOEG.0000031425.72248.85

Sihag, B. S. (2007). Empirical studies on institutions, governance and economic growth: A review. *Humanomics, 23*(4), 204–220. https://doi.org/10.1108/08288660710834694

Weber, M. (1930). *The protestant ethic and the spirit of capitalism*. London: Allen and Unwin.

· 5 ·

ADDRESSING THE RESOURCE CURSE THROUGH AN IDEAL ISLAMIC INSTITUTIONAL FRAMEWORK

Introduction

A truly Islamic economic system is one that is designed to be in compliance with Islamic rules deduced from the Qur'an and the life of the Prophet (sawa). Based on the Islamic vision, an Islamic solution would emphasize and stress amongst others, the following important principles; (1) a greater degree of justice, (2) a higher moral standard, greater honesty and trust to foster mutual cooperation, respect and love for others, and (3) a more equal and efficient distribution of resources and wealth amongst all members of the society. It is against the backdrop of achieving these fundamental principles that the following recommendations for institutional reform are developed.

Proposed Recommendations to Address the Curse and Strengthen Institutions

The recommendations are categorized into three different dimensions which include; (1) Redirection of Education, (2) Duties and Accountability of State Authority and Citizens, and (3) Enhanced supervision. These dimensions are elaborated in further detail below.

Redirection of Education

The ideal society as envisioned by the Metaframework of Islam is made up of believers that are committed to complying with the rules of Allah (swt). At the basic level, the relationship between Muslims and the Creator needs to be truly understood so that Muslims in practice may orient the organization of their daily lives in full understanding of their role. The tainted images of a strict and merciless God, with many forms of punishment to restrict the Muslims are concepts about the Creator that could not be further from the truth and the true essence of Islam. Moreover, this misconstrued and distorted version of Islam has led Muslims to believe that they may engage in corrupt practices, opulent spending, inflict unjust punishment, wage war and seize states and resources.

Mirakhor (2009) explains the most important element of the dynamic relationship between the Creator and the believer is the divine notion of *Wālayah*. Further explanation by Mirakhor on this divine notion is summarized as follows:

1. *Wālayah* carries profound bearings of Love of the Creator for all of His creations and of being in closest proximity with them. It has the capacity to inspire and motivate affection, fondness, attachment and devotion between the lover and the beloved. The consequential reaction allows the believer to feel truly comforted and complete. Active initiation of *wālayah* would on one end, manifest guardianship and authority and on the other, allegiance, loyalty and obedience. What binds these two ends is love and if it weakens, so does the intensity of the *wālayah*.
2. The mission commissioned upon man is to pass on the *wālayah*, the Love of the Creator to His creation. This is stressed further in human love for all humanity. In Mirakhor and Hamid (2009), the authors explain that the ultimate source of all acts of love originates from Allah (swt). They state that love, help, comfort, mercy, devotion, cherishing, and the guardianship and authority that flow from these, in any and all domain or age originate from Allah alone. They further describe that the active-dynamic love of parents for their children is a manifestation of the *wālayah* of Allah (swt).
3. In essence, the active-dynamic love of every creature deserves to be directed ultimately to Allah (swt). Seeking help, deviation, attachment, adoration, loyalty, allegiance in any and all domains should be

oriented ultimately to Allah (swt) alone (Mirakhor & Hamid, 2009). Transgressions, merciless acts towards others would be avoided once one internalizes and realizes that doing so would be in stark violation to the *wālayah* of Allah (swt). The manifestation of one's responsiveness to the *wālayah* of Allah (swt) will be demonstrated by the individual's pledge of allegiance and devotion to the rules of Allah (swt). The cause of corruption and other misdemeanors demonstrates the scant understanding that man has of the *wālayah* of Allah (swt). The misplacement of love towards other worldly preoccupation causes one to deviate from the path of righteousness.

This misplacement can be said to be the ultimate root cause of most of the problems associated with the resource curse. For example, man's love for power results in political patronage, abuse of political power, corruption and rent seeking. Man's love for material wealth and success results in an unbalanced allocation of resources, opulent spending, extravagance and hoarding. The reorientation needs to take form firstly in the direction towards the *wālayah* of Allah (swt). Being cognizant that *wālayah* of Allah (swt) and trusting in Allah (swt) as the ultimate Comforter and Provider will leave little room for deviation. From the teachings of the Qur'an, this comfort provides calm assurance and resolves any apprehension. The power of trusting in Allah (swt) surpasses misdirection that can take place within societies. For example, where it was mentioned before that cultures are to trust only their families and not others is a clear violation of this notion. The unintended consequence to societies as a result of this type of divergence from the true version of Islam would be minimized.

The second redirection is towards the Archetypal Model. All existing practices should be realigned to the Islamic institutional economic framework previously mentioned. The Archetypal Model consisting of the rules governing property rights, institutions of market, production, distribution, and governance represents the fullest possible extent of exemplar society for mankind to replicate. The path for institutional change previously mentioned by North centered on the human quest to invest in new skills and knowledge in order to survive competition.

Along similar lines, the most fundamental path for institutional change to initiate the reformatory and redirection efforts in the two fundamental areas mentioned above lies in investing in knowledge and focusing on the education sector. However, instead of pursuing knowledge for the sake of outdoing the competition, the aim of restructuring the education program is to, as espoused by Mirakhor—"reclaim the vision of Islam." It is pointless to expect Muslims to implement the visions of a true Islamic economy when the teachings on scarcity, property rights and transactions are predominantly based on conventional literature. Cogneau (2003) attributes influence from colonization as the cause for historically western determined institutions shaping language, religion, legal, judicial and political regimes. The author points out that colonization also left a lasting mark on education. For example, the British Empire embarked on concerted efforts to set up a large number of schools despite the two world wars and the 1930s depression.

Government funding was allocated to specifically fund Protestant missionaries to run the schools as British colonization relied heavily on missionaries and traders (Cogneau, 2003). According to the author, the distinctiveness of British policy is to develop education that provided no prospects of power, satisfying a goal to "civilize" and combining missionary actions with government actions. Moreover, the author found that in the case of Africa, the education advantage in former British colonies was not reflected in the countries' economic performance or in the populations' average standard of living. Hence, the conditions of education did not correlate with source of growth and marked improvements in standard of living. Although this result could be linked to the government's own weaknesses, the author explains that the colonial power's identity and own political philosophy do have an influence on the administration.

Investing in a comprehensive early Islamic education program which incorporates Islamic economics may be worthwhile as it is the children that grow up to be working professionals branching into different business or service fields. They should be equipped with sound knowledge on wealth distribution requirements for they would become contributors later on in life. Mirakhor and Hamid (2009) dedicate a number of sections in their book on the conception of knowledge in Islam. They relate the following saying of the Prophet (sawa):

> The seeking of knowledge is obligatory on every Muslim. Ah! Sure Allah loves the seekers of knowledge

The authors' description of Islam and knowledge is provided in further detail below:

- The Qur'an makes clear the importance of knowledge, and it is obligatory for Muslims to consider gaining knowledge as a never-ending pursuit. In a saying, the Prophet (sawa) described three categories as useful branches of knowledge—"the firm Sign (*Al-Ayatu al-Muhkamah*), the Just or Balancing Duty (*al-Faridatu al-Adilah*), and the Established Tradition (al-Sunnatu al-Qaimah)." The word *ayah*, is closely related to mean "sign" in the English form. Firm signs are explained as indicators that provide immediate significance to something they represent. The knowledge of the firm sign is the science of being cognizant of signs that clearly point to the Creator and His relationship to His Creation. It also encompasses being cognizant of the destiny of the creation and humanity. These are explained to be the cosmological dimensions of Islam, and it relates to the cosmological *wālayah* embracing the Love that Allah (swt) gives His Creation His Signs in the universe and humans by which His Creations may achieve cognizance of Him. The adjustment or balancing in the second branch of knowledge refers to justice and balance and encompasses ethical and moral qualities, which it aims to balance. Humans are expected to engage in a balancing duty through encounters with tests, tribulations and trials so that they may experience the significance of behavioral rules as prescribed by the Creator.

- Knowledge of the rules and applying them to one's life are elements described to bring the path from Islam to *iman*. The knowledge of balancing duties relates to the progress of humans, and it is likened to the spiritual form of *wālayah*—that the force of love of the Creator spurs progress, growth and development to humans so that they may strengthen an internal self-command and control that are indispensable to the performance of their duties in this plane of existence.
- The final branch of knowledge refers to the tradition of the Prophet's practices that operationalized the Metaframework in the form of the Archetypal Model. The collection of these traditions is known as the "Sunnah" of the Prophet (sawa) and includes his sayings and practices governing rituals, legal, civil, social, economic and political rules of behavior that explicate, clarify and particularize the general-universal rules of the Metaframework. The main function of the Metaframework is to administer the rules as prescribed by Allah (swt) so that Muslims may orient their daily lives towards achieving excellence in cultivating self-command and control, interaction with others of their kind and the rest of Allah (swt)'s creations and to establish harmony and justice for themselves and for the community. In summary, the rules specify how

humans may adore and serve the Supreme Creator and reciprocate His *wālayah*.

Alam (1996) mentions the importance of having a system of education that is freely accessible based on the Qur'an and Sunnah through the following saying of the Prophet (sawa):

> If anybody goes on his way in search of knowledge, Allah will thereby make easy for him the way to Paradise[1]

Thus, through the branches of knowledge, Allah (swt) equips man with the blueprint design along with methods on how to implement His requirements. One of the reasons the Qur'an was given to mankind was to remove ignorance and darkness and to uphold the rules of Allah (swt). The Islamic education program should be customized to reflect the categories of branches of knowledge so that it can lead humans to a reorientation of knowledge towards the *wālayah* of Allah (swt) and His rules.

Duties and Accountability of State Authority and Citizens

State governance, which includes the duties and accountability of those holding political power and citizens, should be defined and structured according to the Islamic framework. We refer firstly to the work by Kamali (2002) for the discussion on state-citizen relations in Islam. According to Kamali, although the Qur'an and Sunnah lend support to the creation of a political order and leadership that takes charge of community affairs and administers justice, the main actor and audience are still the individual, not the state. The Qur'an addresses the individual and the community of believers when it speaks of the duty of *hisbah*, that is, the "promotion of good and prevention of evil" *al-amr bi'l—maruf wa n-nahy an al-munkar*.

The community of believers is addressed in the Qur'an as "O you who believe" and not as a separate or corporate body. This individual is expected to obey the *ulu al-amr*, which are the persons entrusted with leadership and accountability to the community. The state or any Islamic political organization is service oriented and charged with humanitarian aspects in the sense that the individual remains the principal actor in all its parts. In this respect, the state is not regarded as the supreme political authority; rather it is the *ummah* or the community of believers that is the "locus of political

1. Alam (1996), p. 111.

authority" or those that are said to hold executive sovereignty. This notion of sovereignty is traced back to the Qur'anic doctrine of man as *khalifah* of Allah (swt). It is by virtue of this derived or delegated sovereignty that the community is seen as the repository of political power (Kamali, 2002).

The system of law and government in Islam places ultimate sovereignty only to Allah (swt). Thus, both state and individual are expected to abide by the same law with their duties and rights outlined according to rules of Allah (swt). Together, the objectives of justice, promotion of benefit, prevention of corruption and harm are to be upheld by state and citizen where the former has no authority to overrule or violate the rules of Allah (swt). This harmonious relationship is centered towards achieving the implementation of these rules and is profoundly influenced by the Qur'anic doctrine of unity, *tawhid*.

The measure of the success of the state lies in its ability to enforcing the rules of Allah (swt). Since individual and state are expected to subscribe to the same set of values, and the state exists in order to administer justice, no necessary conflict is assumed to exist between the rights of the individual and the state power (Kamali, 2002). Kamali also provides that the Islamic framework is in contrast with the perception of duality of interests between state and citizen in constitutional law. Islam's perception is instead, inherently individualist as leaders are committed to serve the best interests of the individual. The duty of the state, specifically, is to protect faith, life, property, intellect and lineage through the establishment of a just political order and government (Kamali, 2002).

Mirakhor and Hamid (2009) point out that legitimate recognition may be given to the *ulu āl-amr*, who demonstrate full cognizance of *wālayah*, the adorer and devotee relationship with Allah (swt) and are fully familiar with and adhere to the rules by Allah (swt). In this regard, the authors explain that such persons are such adorer-servants of Allah (swt) that they are considered as worthy to uphold the rulings of (swt). It is the strength of the rule compliance of the authority that gives them the right to oversee the implementation of the prescribed rules of Allah (swt). State authority is not given through other worldly advantages such as riches or military prowess. Citizens, on the other hand choose to follow the commands of the state for the sole reason that they know the authority is cognizant of the orders of Allah (swt) and that they will ensure that rule compliance is strengthened within the community.

The authors provide the example of how the Prophet (sawa) established the accession line of the political authority to depict a rule-compliant legitimate leadership. The leadership was defined as a contract between the person who is deemed worthy of accession to the office and the citizens. Full compliance is given to the terms of the contract for as long as the authority serves as a symbol of operationalization of the rules of Allah (swt). No political authority could retain the position if the rules are violated and on the other side of the coin, no community should remain silent in this situation. Violations of rule compliance, or rather violations of all members of the community to the rule of enjoining good and forbidding evil can also be seen in the existence of corruption, income inequality, oppression and poverty.

The Prophet (sawa) warned against such violations, and Allah (swt) will empower the worst of humans to rule over the community as conveyed in this saying of the Prophet (sawa). Another stated consequence relates to the warning of destruction of communities as a result of non-compliance violation to the rules of Allah (swt). The Qur'an states these examples as the communities during the times of prophets such as Noah, Lut, Hud and Salih.

The study by Robinson et al. (2014) mentioned previously notes that natural resources are typically under state ownership and as a result, decisions relating to natural resource exploitation become national and diplomatic concerns. Alam (1996) covers state responsibility in relation to natural resources and states that it is imperative for management of free resources to be governed by rules of Allah (swt) to minimize exploitation and concentration of wealth. The author points out that one of the reasons for income disparity is the ownership of productive resources. In his view, the rules of Allah (swt) endorse the concept of public property, which implies ownership of certain land or infrastructural assets by the whole community. These assets include seas, roads, bridges, rivers, etc. The same rule applies to every open spring on the land. The ruler cannot claim them for himself or for any particular members of the community. Riches in resources are to be shared and distributed in the most just manner for the benefit of everyone's prosperity, as we have suggested through a national fund with annual checks to all current and future members of the community.

Another role that can be added to the above suggestion relates to the state providing representatives of supervision replicating the market supervisors. According to Alam, the utilization of nature is generally left to the duly constituted state organs to regulate its uses. This regulation should be made responsibly and according to the rules of Allah (swt). It should curb waste, not encourage extravagance and insist that Muslims use nature in the amounts and ways required for production to be converted to justifiable forms for the human needs. The crux of the discussion is that ownership of all resources belongs to Allah (swt) but man has the right of possession to consume, utilize and dispose of the resources according to certain conditions.

In summary, the literature discussed so far provides that both state and citizen are charged with one objective and that is to uphold the rules of Allah (swt). The rules of Allah (swt) for the state authority and citizen are unique, complete and immutable. Its comprehensiveness and completeness represent a system like no other. The rules are all encompassing and direct complete orientation and direction to the fulfillment of man's purpose in the best manner so that it may capture the roles of individuals, states and collective individuals as active members of the society. They are meant to effectively curb corrupt practices, waste, opulent spending, rent seeking, patronage and abusive power plays relating to the institutional issues commonly associated with lax management of resources. They serve to ensure that every member of society experiences a greater degree of justice, a higher moral standard, greater honesty and trust to foster mutual cooperation, respect and love for others and a more just and efficient distribution of resources and wealth amongst all members of society. In situations where citizens are oppressed by unjust practices of the rulers, then Islamic principles dictate that citizens are supposed to actively seek the "promotion of good and prevention of evil" *al-amr bi'l—maruf wa n-nahy an al-munkar*, for example by way of peaceful political revolution and not allow unjust practices to exist for years and years without action to protect the interests of citizens.

Enhanced Supervision

The notion of establishing an independent party who holds a higher authority overseeing the conduct of resource allocation was raised by Acemoglu et al. (2005). The purpose is to limit transgressions by any parties and provide checks and balances. This notion of supervision is not new in Islam and can be traced back during the historical Medinan period of the Prophet (sawa). The Prophet (sawa) created several means for enhanced supervision to oversee the market practices of Medina (Mirakhor & Hamid, 2009). Intense supervision took the form of *muhtasibs* (market supervisor) making their presence felt in the marketplace to curb any misdealing.

This structure can be replicated for the management of resources. Resources that are typically state owned are often given wide-ranging privileges, specifically if some of the board members of the entities are government ministers. For example, in the case of Malaysia, Victor, Hults, and Thurber

(2011) report that the government created the national corporate entity Petronas, which is not required to file annual accounts like all other registered corporations registered in the country. Similarly, in Algeria, the authors state that a conflict of interests arises in the management of the state-owned oil company as the policy function of the country and strategic decision makers of the firm are driven by the same group of people.

Enhanced supervision is especially required to strengthen accountability, supervision and transparency within such politically controlled situations. Resource-rich countries could establish a national level *muhtasib* authority made up of not just the ruling party but members of the opposition to oversee matters relating to resource exploitation and allocation nationwide. The national level *muhtasib* would operate under the full mandate of Islamic rules and teachings to manage oil resources according to the distribution and redistribution mechanisms previously mentioned. The national level *muhtasib* could also strive to ensure all of the country's reporting in relation to resource revenues. It also becomes a platform for collective action to "enjoin good and forbid evil" as no ruling power should be left to conduct the affairs of millions of individuals without undue supervision, especially if the ruling power engages in years and years of unjust oppression, corrupt practices and other unscrupulous acts towards its citizens.

Finally, additional checks and balances may be created by way of channeling oil revenues into a special fund to be managed independently, aside from regular or conventional government budget spending. We refer to the suggestions made by Askari et al. (2006) in establishing a fund for all generations by oil-producing countries. One hundred percent of oil revenues are channeled into this fund, and annual payouts (constant real purchasing power) are made to all eligible citizens of every generation. The authors provide that establishing such fund would involve managing certain implementation issues. The beneficiary needs to be determined and the authors suggest that payouts are made to only those over 18 years of age.

Conditionality of attaining educational proficiency is imposed to these citizens, to ensure that moral hazards associated with wastage and irresponsible behavior are minimized. In terms of management of the fund, an independent entity would oversee the investment activities and ensure that the portfolio is composed of diversified mix of real and financial investments across a broad range of countries, currencies, asset class and non-oil industries. Fund administration is kept independent, transparent with sound governance and control; the success of the fund is supported by an efficient income tax policy and effective taxation system.

Is Economic Reform Possible in the Absence of Political Reform?

There are two undeniable facts about most Muslim countries: they have had sub-par economic performance and they need fundamental economic reforms if they are to become "Muslim Tigers," that is, to achieve more rapid and sustainable economic growth. But can fundamental economic reforms be adopted and implemented without political reforms? Our answer is an unequivocal no.

Effective and sound institutions (essentially the rules of the economic game with the aim of reducing business transaction costs to increase business confidence, investment and thus support a thriving private sector) are the required bedrock of a successful economy. The institutional structure of a society is composed of constitutions, laws and rules that govern the society, its government, its finances, economics and politics; written rules, codes, and agreements that govern contractual relations and exchange and trade relationships; and social norms governing human behavior.

For economic success, conventional economics recommends institutions include a legal system that is fair and just, where all citizens are equal before the law; sound business regulations, their monitoring and enforcement; the protection of private property; contract enforcement; a tax system that is accepted as fair and is universally enforced; and transparency in public decision-making. The Islamic economic and financial system embraces these recommendations but demands even more of Muslims—social and economic justice and morality in all economic interactions. In Islam, the rules, institutions, operations and practice of the Islamic economic and financial system are outlined in the Qur'an and interpreted and put into practice by the Prophet (sawa). While the Quranic outline is immutable, its practice must be adjusted to *prevailing conditions*. The Qur'an provides Muslim societies with the road map to develop effective economic institutions for achieving just and thriving communities.

The Islamic economic system is a far cry from the capitalist system. A comparison of the two reveals some common features, such as private property, profit motive and reliance on markets. Even the economy envisioned by Adam Smith in which the Almighty prescribes rules of behavior—which humans translate into moral rules that govern the behavior of market participants and where self-control, sympathy for others and just behavior limit greed—could be considered a rough approximation of the Islamic vision of an economy. However, since the nineteenth century, capitalism has not resembled either of these visions. It has practices that do not exist in Islam or are prohibited, such as unlimited accumulation of private property and wealth, acceptance of poverty, consumerism and wastefulness, extravagant and opulent consumption, highly unequal distribution of income and wealth, the adverse impact of environmental degradation, and growing financial and economic exclusion.

The hallmark of a rule-abiding Muslim community is justice. An Islamic economy is one where everyone who is able works hard, using knowledge to combine with their own labor and the resources provided by the Creator, to produce goods and services for society. Economic, social and political affairs are conducted with the goal of removing barriers to the progress of all humans. Knowing that they are responsible and accountable, individually and collectively, Muslims invest allegiance in a *legitimate authority* to carry out their affairs, with the legitimacy of the authority established by rule compliance. The Islamic rule, "*commanding the good and forbidding evil*," applicable to individuals and society, assures the full and active participation of all in the affairs of society and a just and representative political system.

The main reasons for the economic under-performance of Muslim countries over the past decades has been non-compliance with the rules of behavior—by Muslims and their governments—and the absence of effective institutions to support thriving societies. How can conditions in Muslim countries be turned around? It requires commitment on the part of individual Muslims to uphold rules, to demand the same of their rulers/governments, and ultimately to develop and nurture the needed institutions.

Muslims must individually and collectively take charge of their religion and openly debate and question what their religion says about rules, institutions and the outcomes that they should expect. They should discuss what their religion indicates in practical terms, for the political, social and economic characters of a successful Muslim society. Memorizing the Qur'an, while useful in itself, is not a substitute for understanding the Qur'an and its interpretation by the Prophet (sawa).

Moreover, Muslims should not rely on individuals with agendas—dictators, extremists, selfish rulers, politicians or clerics—to tell them what they should and should not do as "good" Muslims. God has given humanity the freedom to choose and Muslims should take advantage of this invaluable gift. The Qur'an requires individual Muslims to develop themselves, the physical world and their societies. Individual Muslims have a critical role and must do their part in creating just and flourishing communities. At the same time, they must hold their leaders accountable and answerable to the community.

We started this section with the question—can fundamental economic reform that is essentially made up of effective institutions be adopted and implemented in the absence of political reform? Our answer was no. After considering what reforms and institutions entail, the reasons for our answer should be evident. Dictatorships, absolute monarchies, military rule and theocracies that justify their rule by invoking God's will not secede the power that has enriched them and has facilitated the continuation of their corrupt and unrepresentative rule. Simply said, under the rule of law and representative governance that are basic institutional requirements, most Muslim rulers would be swept away.

Anyone with eyes can see that most rulers in the Muslim World are not rule compliant and have used religion for control and self-enrichment. It is not Islam that threatens the non-Muslim world, but its misrepresentation and Muslim dictators whose selfishness and autocratic rule have fueled extremism and millions of disenfranchised Muslims.

Dictatorships, autocratic rule and economic failure and their unfounded association to Islam are an important cause of extremism, terrorism and instability. It is better institutions and meaningful economic and political reforms in Muslim countries that are essential for development and progress and peaceful co-existence with the West. But the needed institutions, especially the rule of law and representative governance, will not develop under existing political conditions in most Muslim countries. For the required economic institutions to develop, be nurtured and be effective, they *must* be accompanied by fundamental political change.

Conclusion

The most frequent suggestion in the conventional literature to address institutional weaknesses and the resource curse is decentralization as the issue of the resource curse is tackled in parts. Where some of the literature provides detailed descriptions of tackling government malfeasances, issues pertaining to resource allocation are left unaddressed. The policy recommendation advocated according to Islamic teachings provides a blending of insights into restoring social, economic and political justice. Through the redirection of education, the reclaiming of Islamic vision is proposed. Individual Muslims should study the Qur'an and the life of the Prophet (sawa) more closely, debate Qur'anic teachings and its practice by the Prophet (sawa), and adopt the indicated rules and their applicability to prevailing global conditions.

The Muslim communities should demand rule compliance on the part of their leaders and governments. Unless Muslims take back their religion from those who use it as an instrument of control and power for personal gains, there is little hope that effective institution building will take hold. The redirection of education towards the *wālayah* and rules of Allah (swt) will set the path for cognizance and realization on how economic affairs pertaining to resources of the country should be conducted.

From the Islamic viewpoint, social, economic and political prosperity are achieved through actions of shared responsibility by state and citizens, especially by individual citizens. Every member of the community is accountable to curb actions that are in violation of the rules of Allah (swt). Finally, through enhanced supervision, rule compliance is enforced to ensure social order and justice are upheld for all members of the community.

References

Acemoglu, D., Johnson, S., & Robinson, J. a. (2005). Chapter 6: Institutions as a fundamental cause of long-run growth. *Handbook of Economic Growth, 1*(SUPPL. PART A), 385–472. https://doi.org/10.1016/S1574-0684(05)01006-3

Alam, D. M. M. (1996). *Perspective on Islamic Economics* (1st ed.). New Delhi: Institute of Objective Studies.

Askari, H., Abbas, F., Jabbour, G., & Kwon, D. (2006). An economic manifesto for the oil exporting countries of the Persian Gulf. *Banca Nazionale del Lavoro Quarterly Review*, 59(239), 363–388. Retrieved from http://ovidsp.ovid.com/ovidweb.cgi?T=JS&CSC=Y&NEWS=N&PAGE=fulltext&D=econ&AN=0901175

Cogneau, D. (2003). Colonisation, School and Development in Africa: An Empirical Analysis. DIAL Document de Travail DT/2003/01. DIAL

Kamali, M. H. (2002). *The dignity of man: An Islamic perspective.* Selangor: Ilmiah Publishers Sdn Bhd.

Mirakhor, A. (2009). Islamic economics and finance: An institutional perspective. *IIUM Journal of Economics and Management, 17*(1), 21.

Mirakhor, A., & Hamid, I. S. (2009). *Islam and development.* New York, NY: Global Scholarly Publication.

Robinson, J. a., Torvik, R., & Verdier, T. (2014). Political foundations of the resource curse: A simplification and a comment. *Journal of Development Economics, 106*, 194–198. https://doi.org/10.1016/j.jdeveco.2013.09.004

Victor, D. G., Hults, D. R., & Thurber, M. C. (2011). *Oil and governance: State owned enteprises and the world energy supply.* Cambridge: Cambridge University Press.

· 6 ·

POLICY ASSESSMENT

Saudi Arabia, Malaysia and Qatar

Introduction

We conclude that the OIC oil-producing countries in our study have conducted their affairs in a fool's world—as if oil prices would stay above $100 per barrel for all time (with little competition from other producers and renewable sources of energy) and that their oil reserves would last forever! In 2016, some may have been awakened from their slumber. As they look ahead, the more populated Muslim countries among the major producers, such as Saudi Arabia and Nigeria, see unsustainable budget deficits, economic stagnation, high unemployment and potentially catastrophic domestic conflict and upheaval. We again restate that the grim economic condition of the OIC oil-producing countries shows that there is now an urgency, more than ever to assess and evaluate current weaknesses in their developmental policies, which require effective institutions at their foundation.

In our previous chapter, we provided recommendations for Islamic institutional reform. Our purpose here is to narrow down and select three countries (Saudi Arabia, Malaysia and Qatar) from our countries in the study and provide a more comprehensive explanation on how these countries have failed and succeeded and how can our recommendations be implemented. In order to do so, we take a more in-depth study on the economic, social and political effects of oil for these three countries. We examine in greater detail the reasons for the failure in their quality of institutions, largely highlighting the extent they have stifled development and growth. We conclude the chapter by providing customized recommendations for these countries to adopt as a way forward in order to achieve economic and social justice for their current and future generations.

Generally speaking, when it comes to the management of depletable natural resources, the recommendations of conventional economics and Islam are similar though still different. Conventional economics recognizes the fact that oil and gas reserves are a part of a society's capital endowment and should not be simply depleted. Instead they should be developed optimally, replaced by alternative forms of capital to benefit future generations. Islam recognizes such natural resources as the Almighty's gift to all humanity of all generations. It belongs to no one but to the Almighty because he created it. It is to be used on this earth to benefit all individuals of all generations equally. The promotion of justice must be at its foundation. Clearly, it cannot be depleted to finance one person's, one country's or one generation's consumption. It must benefit individuals equally with an eye to affording the same to future generations. However, realistically in a world of sovereign nations, we would argue that this sharing of benefits should be limited to the nation state where the resource is located until humanity becomes one and sharing rules supreme.

The task of Muslim oil-producing countries is clear but difficult. First, governments must take control of all minerals. Second, governments must make sure that they do not waste depleting mineral resources, because they are the birthright of all citizens and must be used productively. Third, as minerals are depleted, governments must make sure that they use their revenues in such a way that all citizens today and for all time receive similar real benefits. The application of equity in the depletion of a depletable resource is a perfect example of how Islamic teachings must be practiced by any legitimate government in a Muslim country. Oil should benefit all members of the current generation equally (with optimal depletion and optimal compensation for depletion), with the implication of relatively even distribution of income given the overwhelming role of oil (as opposed to hard work and sound productive investments) in some of these economies (such as Abu Dhabi in the UAE, Qatar and Kuwait), and that these benefits should be similar for *all* generations to come.

The equitable, but not equal, distribution of resources and output in Islam does *not* allow affluence and poverty to coexist side by side. Islamic view emphasizes absolute scarcity not being the root of distributional controversies. Relative scarcity, discrepancy between resources and wants, is the central issue, since scarcity drives its significance from human wants, which are unfulfilled. Equity and efficiency considerations in Islam are a result of the principles governing economic activity in general, and, in particular, principles pertaining to the ownership and utilization of resources, and the distribution of the final output. Natural resources are to be shared equally by all, either in their natural form or through their contributions to output. Man may acquire a priority of use for some natural resources and is allowed to keep them as long as he is fulfilling his obligations towards society, the initial owner of these resources.

As we have outlined earlier, many countries and states (within countries, such as Alaska) have adopted funds for future generations. These funds generally take but a *small* percentage of revenues of exhaustible resource depletion; their operations in some countries are opaque; and in some cases there are no rules for the portion to be saved; in cases where there are rules, the rules for the portion to be saved, its management, and/or its distribution are highly imprecise. These funds do not follow Solow's prescription of intergenerational equity, as they do not replace all of the drawn-down resource with future sources of output. Most egregiously in many of the Muslim oil-producing countries, ruling families or the elite takes a significant portion of the revenues even *before* it becomes available for public expenditures or for placement in any sort of a fund, whether for revenue stabilization or for future generations.

The average citizen has been marginalized in most, if not all, Muslim oil-exporting countries. There is thus a vicious circle of economic failure, corruption and popular anger. Our proposition is that the only path of progress against such odds is to take *all* oil and gas revenues away from the state and give them directly to citizens in a manner that is intergenerationally equitable. The *de-linking* of oil revenues from government coffers may avoid some other problems normally associated with the exploitation of depletable natural resources. As we have mentioned above, easy oil revenues tend to fuel military expenditures. Military expenditures, in turn, could be associated with a number of other adverse developments. Civil wars and conflict are more likely if military expenditures are high; and the risk of civil war is higher if natural resource endowment is double the average.[1] Civil wars, in turn, lead to higher military expenditures, capital flight, loss of social capital, slower economic

1. See Collier et al. (2003).

growth, and more poverty and refugees, an almost impenetrable vicious circle. The adverse role of natural resources is generally accepted.

Some advocate using all the proceeds from natural resource depletion for establishing diversified economies. This is easier said than done. Corruption invariably raises its ugly head. Governments invariably fail to make good investment decisions. The development of a diversified economy today invariably affords those well positioned to benefit more than the average citizen unless the government enforces an equitable tax regime to redistribute benefits, a technically difficult task and politically impossible. Others advocate an "international template for the acceptable governance of natural resource revenues to which a government with significant revenues could chose to subscribe."[2] We believe that a fund that takes all revenues away from the government should be an integral and primary component of any template.

Saudi Arabia

Saudi Arabia has received more oil export revenues than any OIC country. Unfortunately, much of this depletable wealth, instead of being transformed into other forms of productive capital, has been wasted over the years. Oil revenues jumped during the 1970s and early 1980s. But then for nearly 20 years Saudi Arabia ran budgetary deficits and had to resort to borrowing. Rent-seeking activities, corruption and waste drained, and have continued to drain, the national treasury. Subsidies, both consumer and input subsidies, were the government's preferred way to deliver the benefits of oil depletion to the populace and to buy domestic political support. Public employment, at salaries well above those in the private sector, slowly became almost an expected Saudi birth right as the private sector could not provide the jobs that Saudis expected. Saudi Arabia for a number of years, since the Iranian Revolution, has been the single largest importer of military hardware.

While military expenditures have been justified on account of the perceived military threat from Iran, they have also been an important instrument to buy foreign support for the rulers, principally from the US and the UK. Saudi Arabia has also used its financial resources to spread its wahhabi version of Islam around the world and to get the backing of Arab strongmen in the region. At the same time, the Al-Sauds, a tribe that allegedly has more than

2. Ibid. p. 182, for a five-point template.

20,000 princes and princesses, have had direct access to the treasury in the form of annual stipends, which for the senior princes runs into tens of millions of dollars and for the ruler has no limit.

As a result of their policies and practices and low oil prices, the country is now facing serious financial difficulties. Yes, the country has some world-class infrastructure, including first-rate hospitals and good universities, but the private sector cannot provide jobs, let alone the high-paying jobs, that the population expects. If the rulers had been forward looking 40, 30 or even 20 years ago, they could have adopted reforms gradually and avoided the present crunch.

After the death of King Abdullah, his successor King Salman appointed his young son Prince Mohammad as the Deputy Crown Prince. Prince Mohammad wasted no time in admitting that the Saudi model—lavish consumer subsidies, uncontrolled expenditures in all areas of welfare and vast military expenditures coupled with little or no taxation—could not be sustained. If he needed further evidence, he could have consulted a recent book that laid bare the fact that a crisis is looming unless drastic action is taken.[3] The young Saudi Prince has embraced a report coined "Vision 2030" prepared by an American consulting firm as his roadmap. Over the next 15 years or so, he intends to transform the Saudi economy to be independent of oil revenues. This is easier said than done.

While Prince Mohammad may have noted a part of the problem at hand, he may have not fully appreciated a number of related issues. First, there is the lavish, and un-Islamic, lifestyle of his father the king and for that matter the lifestyle of his large family all financed by oil. But such unsustainable expenditures also carry over into military imports, financial aid for foreigners who back the Al-Sauds, public sector jobs as employment of last resort and into vast inefficient consumer and industrial subsidies. The termination of these wasteful expenditures will take much courage and time. They have become a part and parcel of the Saudi landscape. Second, although he has stated that he intends to channel an increasing percentage of oil revenues into a super sovereign wealth fund (SWF), he has not clarified a number of associated issues. Does the fund ultimately belong to the citizens of Saudi Arabia or will the Al-Sauds have privileged access? What are the objectives of the fund? Will the SWF issue annual checks to all Saudis of all generations? How and under

3. Abdulaziz Al-Dukheil, Saudi Government Revenues and Expenditures: A Financial Crisis in the Making, New York, NY: Palgrave Macmillan, 2013.

what circumstances can its resources be tapped? Will it act as a backup in case of budget shortfalls? How will the fund be managed in order to be transparent and immune from corruption?

There is a clear temptation not to establish objectives, policies and rules for such a fund in order to preserve "flexibility" for the rulers. Third, Vision 2030 will succeed if and only if there is a national dialogue to persuade the general citizenry to buy into the program. Such a dialogue must be free and embody no threats; it must persuade the average Saudi that it is fair and just with every segment of the population, including the ruling family, making sacrifices commensurate with their ability to cut back; and that the program is in lock step with Islamic teachings. We say all this because the transition will be very painful and if the country succeeds in this quest, political liberalization will have to also go hand-in-hand with the program. It is nearly impossible to introduce effective institutions, including the rule of law, transparency and much more and continue to rule as the Al-Sauds have done in the past. To succeed, economic liberalization will have to be accompanied by political liberalization. Are the Al-Sauds ready for such change? What we have said about Saudi Arabia applies, with some modification, to Iran, Iraq and to a number of other Muslim countries.

We have initiated this chapter with Saudi Arabia because the rulers have admitted the need for economic reforms and have a partial plan to move ahead. Other Muslim countries that are complacent about reforms are at their own peril. All natural resource-rich Muslim countries are in desperate need of institutional reform and the sooner they start on this path the easier will be their journey! They should delineate a detailed plan for institutional reform; incorporate a realistic timetable to make their adoption and the transition as smooth as humanly possible; and they should invite public scrutiny so that the general citizenry buys into the plan as a credible commitment. But we must emphasize what we have said already a number of times—meaningful economic reform and effective institutional development cannot take hold unless it is accompanied by political reform. How can an effective legal system (with equality for all before an independent judiciary) be established under absolute monarchies? How can oil revenues, the benefits of natural resource exploitation, be shared equitably with all citizens of all generations when hereditary rulers have direct access to the national treasury? These and many more such dichotomies highlight the internal conflicts and the reasons why progress will be near impossible unless political and economic reform (institution building) go hand-in-hand.

Malaysia

Malaysia joined the OIC in 1969.[4] According to the Malaysian Department of Statistics, the Muslim population of the country is 61.3% of 28.3 million total population. The ethnic group breakdown for Malaysian citizens is provided in Figure 6.1.

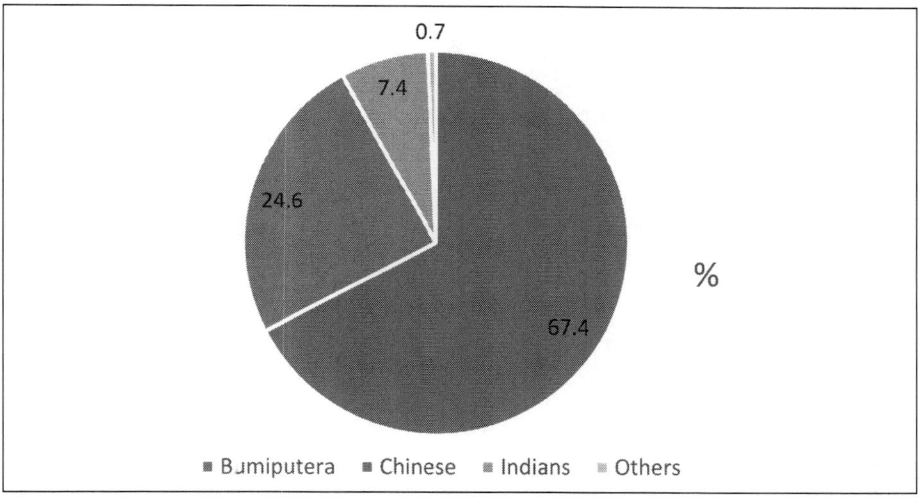

Figure 6.1: Percentage Distribution of the Population by Ethnic Group, Malaysia.
Source: www.statistics.gov.my

The Malays, or local indigenous formally known as the Bumiputeras, constitute the predominant ethnic group in Malaysia. Chinese and Indians make up respectively 24.6% and 7.4% of the total population. The multi-racial society in Malaya (pre-independence Malaysia) developed in the 15th century when an influx of migration concentrated in Malacca, the country's trading hub between India and China. The arrival of immigrants continued more rapidly in the middle 19th century, and it was at this time that the predominantly Malay society changed to become multi-racial as Malacca's geo-strategic trade location gained prominence more rapidly at the start and middle of the 19th century. Many traders from China, the Middle East, Java and Sumatra began to settle down and migrate to Malacca (Abdul Khalid, 2014). Another great influence on the country's racial and economic landscape was from the British during their colonial period from the years 1886 until 1957.

4. http://www.oicun.org/3/28/2016

According to Abdul Khalid (2014) the beginning of large-scale immigration of Chinese to Malaya at that time coincided with the increase in demand for tin in the industrial countries of Europe and the United States, which then led to a marked influx of Chinese labor and capital. Abdul Khalid explains that the Indian immigrants were recruited during the rubber boom period after 1905 due to the expansion of the motor industry led by Henry Ford. It was the colonial government and rule that prompted this migration as they targeted specific groups of the male population and those from the lower income class. Khalid states that these groups were favored by the British because they were considered more subservient laborers than others.

As the population grew, education system and political organisations were set up to protect the Chinese and Indian interests in the country. The relationship between the Malay and the Chinese, although non-violent, was not harmonious and was plagued with unspeakable tension. It was also during this period, the Malays sought to distinguish themselves apart and maintain Malaya, first and foremost as a Malay nation. The Malays worked with the British and formulated a constitution that would almost always preserve special Malay privileges. Each of the 11 states in Malaya were governed by a Malay king or ruler, whom according to historians governed by way of special superiority and acknowledgement was to be given to their royal lineage. To this day, the customary obeisance ritual, through the bowing down of one's head and clasping both hands on the forehead as if in prayer is practiced when one meets the king. Aimed to demonstrate a high order of reverence for the king, this ritual is meant to segregate the races and distinguish between those with noble blood and those born without.

The British acknowledged the majority Malay rights in order to protect their trade interest in the region. According to Abdul Khalid, the earnings from tin and rubber were so important that the British termed Malaya as the "dollar arsenal of the Empire." He reports that British Malaya was the world's largest rubber producer and the Empire's largest hard-currency earner after World War I when Britain was cash strapped and recovering from post-war effects.

When the constitution for independence from British rule was being developed between the Malays and the British, Shamsul (2001) reports that in order for the Chinese and Indians to effectively became citizens they had to acknowledge Malay dominance, a special privilege given to only the Malays to be appointed as kings and rulers, in education and government services and recognise Malay and Islam, respectively, as the official language and religion for the new nation-state. Shamsul further provides that it was in 1963, during the formation of the Federation of Malaysia, that the definition of "Malay" included a new dimension and that was to include local groups in the states Sabah and Sarawak from East Malaysia. East Malaysia consisted of Muslim and non-Muslim tribal groups. In order to gain a higher percentage of electoral seats in the local legislative assemblies of Sabah and Sarawak, the federal government used the term Bumiputeras (son of the soil) to accommodate the Malays and the native Muslims and non-Muslims in a single category (Shamsul, 2001).

This colony that the British created marked racial territories in its trade activities. The more profitable and productive sectors were dominated by non-Malays and the Malays were in low-level, low-producing occupations such as peasant agriculture (Abdul Khalid, 2014). Chinese business activities were concentrated in tin and rubber production, haulage, transportation, light manufacturing and in distributive and service trades (Koon, 1996; Machado, 1998). Geographical segregation was also in part due to the economic specialisation as the tin mines were in the urban area, rubber plantations were semi-rural and the agricultural activities were predominantly in the rural areas. Machado states that the communities were kept apart and there was limited interaction or integration between them.

Dissent started to grow among the Malays as the income gap grew, and the Malay group's education and material progress became the most backward. Additionally, locally born Chinese started to request rights similar to native Malays. Abdul Khalid (2014) reports the following response to the demand of the Chinese by Dato' Onn Jaafar, leader of the Malays:

> We know very well that there is not only one race staying in Malaya, that there are also other Asiatic races inhabiting this land. We have kindly allowed them to stay here; but now they have even started to clamour [for] our land. But notwithstanding all their attempts, we shall firmly maintain this land belongs to us. I speak so because I notice that the other race in the peninsula affirms that they have the right to demand equality of treatment. Whatever comes, we will never consent to this. We wish to make them understand that they are only tenants of the house but not the owners ...[5]

Dato' Onn[6] proceeded to form a predominantly Malay party called the United Malay National Organisation (UMNO) and unity or compromise was

5. Abdul Khalid (2014, p. 65).
6. Malay names indicate that a person should be addressed by the given name, as the surname is a patronym and not a family name.

made with the Chinese and Indians when the British insisted racial unity was a condition for independence. The new alliance, called the Barisan Nasional largely consists of Malay dominance, with special representation given to the Chinese and Indians provided that the non-Malays recognize special privileges of the Malays "without any limit of time" (Abdul Khalid, 2014).

The resentment that arose from the colonial-instilled rule incorporating racial and economic discrimination culminated in a socio-economic imbalance among the different ethnic groups and consequently led to racial riots in 1969. Protestors took to the streets and the Chinese taunted the Malays to "go back to the villages as we, the Chinese are now in power" (Ahmad, 2007). The clash engulfed the city of Kuala Lumpur in racial killings, street fights and vandalism. Some 6,000 residents, of which 90% whom were Chinese lost their property and homes (Koon, 1996). The Chinese defeat, according to Koon (1996) revealed the undisputable Malay superior power backed by Malay-controlled military and police force.

When order was restored, a new 20-year economic policy (NEP) was formed to address economic inequalities. It was meant to reduce the poverty rate among the Malays, giving Bumiputeras a 30% share of corporate assets, a minimum of 30% participation in employment and private companies, special privileges in government contracts, projects, financial assistance urban housing and scholarships. Koon provides that UMNO initiated legislative and executive measures establishing state agencies and bodies by the hundreds in order to acquire economic assets on behalf of Malays. He adds that the policy proved to be a success when it ended in 1991 when official figures showed a poverty rate decrease from 49.3% in 1970 to 15% in 1990. The Malays were reported to have left the agriculture sector where 30.5% were employed in secondary (mining, manufacturing, construction and transport) and 40.5% in tertiary (wholesale and retail trade, finance, government and other services) sectors.

While the NEP could be considered a success and assisted Bumiputeras to gain limited economic strength, the income gap remains between the races, and the targets developed to help the Malays have not all been met. In order to expand economic control the Chinese sought to retain and increase their economic wealth by deploying the "NEP by-pass strategies" (Abdul Khalid, 2014). According to Koon (1996) the Chinese had superior access to capital and credit through their associations, which served as networks for members to exchange information on market conditions and sources to expand business. In addition, their strong network of connections led them to form alliances with top political officials who awarded projects in their favour for economic and political support. Koon states that the NEP bypass saw the Chinese developing private colleges that ran twinning programmes with foreign universities.

Tension with the Chinese dramatically widened preferential policies to maximise Malay opportunities in every sphere of social and economic activity, including in the oil industry. In the mid-1970s, specifically after the oil prices soared in 1973/1974, petroleum production began providentially in the country off the east coast of Peninsular Malaysia (Khan, 2000). It was during this period that Malaysia became a net exporter of oil. Resources in the pre-colonial period were state government prerogatives; however, in 1974 the Petroleum Development Act gave federal authorities jurisdiction and exclusive centralised rights over petroleum resources.

Khan states that while petroleum royalties are shared with the state government, the federal government manages and controls Petronas, the National Petroleum Corporation, and all other petroleum revenues. Additionally, Petronas was given exclusive rights for exploration and exploitation on both onshore and offshore petroleum reserves. Since its inception in 1974 and to this day, the Prime Minister alone selects the Chairman of Petronas.

According to Gale (1981), the Prime Minister had said that the appointment of the first Chairman of Petronas Tengku Razaleigh, Hamzah was equivalent to that of cabinet minister, thus enforcing his personal interest and support for the organization. Gale reports Tengku Razaleigh was formidable in pursuing the Bumiputera agenda as he provided an indefinable sense of economic leadership to the local business communities and resented the domination of the Chinese in the national economy. Tengku Razaleigh resigned from his position in order to become Finance Minister. Gale provides that in losing a dynamic and influential Chairman, Petronas showed signs of lethargy in developing new policies or plan for further expansion on its own initiative and often acted only after being given government instruction. He adds that after 1976, the implementation of the National Depletion Policy[7] and the directive for Petronas to expand into the local retail market were given directly from the Prime Minister. Herein, lies the dilemma of Petronas—it is

7. Policy to safeguard the exploitation of natural oil reserves due to the rapid increase in the production of crude oil (Jalal & Bodger, 2009).

obliged to operate in accordance to national interest, however, in doing so, conflicts often arise with the interests of foreign oil companies, and it serves to accommodate them with the requirements of the nation's political and economic realities (Gale, 1981).

A few other government assignments were given to Petronas in the mid-1980s—it was forced to finance a commercially unviable mega property city-development projects and in early 1990s it was directed to "rescue" the Bumiputra Bank on several occasions (Khan, 2000). Petronas became an instrument to use state resources to finance economically problematic and politically inspired projects even under the administration of Prime Minister Mahathir Mohamad from the years 1981 to 2003. Machado (1998) wrote that at the time of Mahathir's regime, the government achieved great executive power and even made amendments to the Federal Constitution. Through exercise of power and executive dominance, the state-created rents were channelled to well-connected businessmen and implementation of key economic policies involved political patronage in determining the access and allocation of rents. Machado views Malaysia's impressive economic growth in spite of misallocation of resources as paradoxical because state-created rents were dissipated by the rent-seeking activities that followed them.

Government deregulation activities lead to privatisation of infrastructure development, which includes a new international airport, highways, improved port facilities, major telecommunications projects and increased energy supply. The middle class started to grow with lower unemployment, increase in real wages and increased business opportunities. Machado claims that executive dominance enhanced the government's development capacity on the one hand, and political patronage as well as abuse of power on the other. He adds that projects were awarded to favored individuals, and with the increased state power through implementing policies, regulating markets and granting of contracts, access to rents was contingent on political access and influence. Machado claims that the political elite abused their power to capture the rents for themselves, and public knowledge caused the party to make their activities even more covert.

Machado explains that as a result, new modes of rent allocation involved favoring businessmen who were politicians or politically well-connected at the cost of having these rents better deployed to accelerate growth and generate productivity gains. Those who were willing to provide economic and political support received state rents by the millions in their favor, thus widening the income gap and undermining efforts to properly allocate resources.

To provide an example of political patronage that took place during this period we refer to the work of Ebrahim et al. (2014). The authors indicate that it was the economic, social and personal interest of Mahathir that led him to set up Hicom, one of the largest industrial firms in Malaysia. It did not change, when his successors Abdullah Ahmad Badawi and Prime Minister Najib Abdul Razak took over. At the same time, three other national policies have been implemented after the NEP. The National Development Policy (NDP) from 1991 to 2000 focused on ensuring balanced development of major sectors of the economy and still targeted reducing socio-economic inequalities across communities, According to Ragayah (2008) the strategy of growth was still with equity and also aimed at reducing inequalities within regions and among Malaysians. The author adds that the private sector was more relied upon than the public sector, whose role was complementary and just as supportive.

The National Vision Policy from 2001 and 2010 focused on nation building and resilience and the National Transformation Policy currently in effect from 2011 and running until 2020 focuses on raising GDP, levels of household income and public well-being.[8] It consolidates the two previous policies to attain a united, progressive and prosperous society. While some have contended that the policies succeeded in creating a more diversified economy as Malaysia is now a leading exporter of electronics, palm oil and natural gas, others report that the NEP and the NDP encouraged political patronage (see Ebrahim et al., 2014). Ragayah found that inequality has widened during the NDP and NVP periods and warned of its negative effects on social cohesion, political stability and translating to inequality of opportunity across generations.

The preceding paragraphs have established the background to discuss the following key challenges that Malaysia currently faces in the management of resources:

1. Wealth inequality among the races—we see that the British instilled a racially inspired economic trade-off resulting in socio-economic imbalance among the races. As a result, government intervention in economic policies and resource management has always been along ethnic lines. Perceiving that the government did not cater to the

8. Taken from the Malaysia Plan (2016–2020).

Chinese, this group proceeded to establish networks and associations that would exclusively only help members of their community.

Discrimination in the workplace also continues to this day as the Chinese are discriminated in the public sector and the Malays in the private. However, Abdul Khalid (2014) claims that the public sector only makes up one-tenth of employment in the labor sector. He believes that discrimination in the private sector by the Chinese, which makes up about 85% of employment, is more prominent as preferential treatment is rampant, in the form of lower salary, delayed promotions and limited access to training, even denying employment to non-Chinese. Abdul Khalid found through random checks on job advertisements, employment requiring either Chinese or Mandarin proficiency is not uncommon even though jobs were catering to the domestic markets and being conversant in Mandarin provides no added value.

Abdul Khalid also believes that because the Chinese had a head start in wealth accumulation during the colonial period, Chinese families were able to accumulate wealth and pass it down to future generations. He reports that the average Chinese in 2009 at an early age of 25 had 2.5 times more wealth than the Bumiputeras. In another study,[9] 44% of children born to Chinese parents with no formal education were found to have tertiary education while for the Bumiputeras it is 33% and Indians, just 5%. The same study found 39% of Chinese children born to low-skilled parents are high skilled while for Bumiputeras the rate is 25% and Indians 19%. *Forbes*'[10] 2016 Malaysian billionaire ranking is made up of predominantly Chinese, indicating that the status of economic wealth and power has not changed to this day.

2. Poor governance—the ruling party's patronage and corruption activities are widely known with one former Prime Minister claiming that the supreme council members and ministers are mainly only interested in what they can gain out of politics for their own selves (Abdul Khalid, 2014). Rampant political lobbying and activities undermine competitive market forces and benefit only selected groups. These groups, often proceed to maintain their wealth and power and as a

9. KRI (2016).
10. "Malaysia's 50 Richest" (2016).

result interventions may not be easily implemented. According to Fraser (2006), the Malaysian economy has come to be known as relationship capitalism and that by 1995 almost 20% of UMNO's division chairmen were millionaire businessmen. Why then are the Bumiputeras still left behind in their share of the nation's wealth?

In our view, the benefits gained by selected Malays were not channelled to other members of the society and continue to circulate within the elite ruling party group. We refer to the research conducted by Caselli and Cunningham (2009) for the explanation of how the behaviors of political elites could generate a resource curse. The authors first distinguish between centralized and decentralized political mechanisms. Centralized mechanisms are those pertaining to the constraints and incentives faced by the political elite, whose main concerns as direct recipient of the resource curse include deciding on how to allocate resources between their own enrichment, increasing the elite's chances of retaining power and investments that would generate non-resource income.

Centralized mechanisms, are a direct fit to authoritarian regimes. In it, resource abundance directly affects decisions made by the elite as the resultant effect is an increase in the value of staying in power. The authors provide two broad scenarios in which staying in power can lead to a resource curse. The first, involves the leader facing limits in budget and time. When the return to stay in power is great, allocation of resources moves away from productive activities and is channelled instead to preserve him in power. The second, relates to the spill-over effects to the private sector from patronage activities. When the leader is allocating spending in unproductive activities such as military and security apparatus, it allows him to be in office for a longer period and increases the changes of spending more on productive activities in the broader time horizon. In such a case, patronage-led activities turn resources into a blessing. If, however, resource windfall relaxes government's constraints and increase in resource revenue lowers the leader's marginal utility of consumption, leaders often indulge in leisure. If leisure, comes at the expense of devoting energy and time towards productive policy making, then it is possible to generate a curse.

Decentralized mechanisms refer to rent seeking where resource rents change the incentive structure for individuals, causing them

> to reallocate effort from productive to unproductive sectors. In the setting of exploiting reserves under government owned companies, transparency, access to capital, freedom to retain profits and patronage-driven activities all depend on the political equilibrium. The extent of resource curse activities is related to the behaviors of the political leader. The study found that leaders who were operating under constraints in budget and time and caused substitution away from productive activities resulted in a curse. Specifically, the incumbent keeps political institutions weak.

In summary, the need to establish a separate fund for oil revenues has become even more critical for Malaysia. Such a fund, should be managed transparently with an independent entity to oversee its activities. Any government bailouts or borrowings made from the fund should be declared in the most transparent manner. The objective of the fund, would be to safeguard the interests of the public by ensuring just and fair allocation of resources. Management of the fund should be made in accordance to Islamic principles and teachings because its application provides no tolerance for selective preferential treatments, handouts, misappropriation and corrupt dealings. In limiting the control of any ruling party, a term condition could be legislated, limiting any ruling party or individuals to hold office for no more than two terms. Declaration of assets by ministers along with their family members should also be made mandatory and a condition for reappointment.

Efforts to eradicate racial tensions should be made a priority as an Islamic framework views humankind as one *ummah*. Redirection of education should be made towards realising one of the most important principles of Islam relating to the Oneness and Uniqueness of the Creator, which precipitates the unity of creation and mankind. The believers of Islamic teachings see only themselves when looking at others, as the Qur'an provides clear guidance that humans are fundamentally one unit.

According to Mirakhor (2009) humans were created of one self, and will return to their Creator ultimately as one self; "Neither your creation, nor your resurrection will be other than as one united self," Qur'an Surah Ali-Imran, Verse 28. Mirakhor adds that it is this unity that reflects the Unity of the Creator and leads humans to extend *wālayah* towards one another. The loving adoration towards the Creator represents a common compass to extend loving care to one another as all humans are brothers in humanity and are related to one another, ultimately sharing one father and mother. They should share one common objective and a common enemy. As such, according to the ordains of Allah (swt), love for other humans should be made central to a believer's life and the resulting effect leads to nearness to the Creator.

Qatar

The Al-Thani tribe rules Qatar, but the people of Qatar are more fortunate than their neighbors in Saudi Arabia for three main reasons—the oil/gas endowment per capita of Qatar is multiples of that in Saudi Arabia, the Al-Thanis appear more open-minded than the Al-Sauds and a significant percentage (5%) of Qatar's native population of roughly 200,000 are in fact Al-Thanis. Qataris enjoy the highest GDP per capita in the world (at well over $100,00 per capita of Qatari citizens) and above average economic growth among Muslim oil exporters; if the Qatar's natural gas reserves are valued very conservatively, the per capita wealth of Qatar would run into the millions of dollars. As a result, Qatar has not faced the economic pressures and outlook confronting Saudi Arabia and Malaysia. The emirate of Abu Dhabi (a member of the UAE) is in a similar position as Qatar. As a result, the Al-Thanis may have more time to establish effective institutions, but they have to start before different Al-Thani family factions (there are basically four major branches in the Al-Thani tribe) stage a takeover or the general citizenry demand change.

Conclusion

Our country assessment provides further confirmation that countries with poor institutional quality are unable to maintain socio-economic justice and fair distribution of resources. We see a more in-depth scale of corrosion in quality of institutions being a result of conducting practices that are a departure from Islamic principles and teachings.

In the case of Saudi Arabia, the economic landscape is defined by opulent spending (including consumer subsidies to buy support and outrageous lifestyle of the rulers), waste (unnecessary military expenditures in part to buy foreign support and corrupt practices) and misallocation of resources (showcase expenditures as opposed to infrastructure to serve the needs of the people), with little to support sustained economic growth as oil is depleted. The new

King and his Deputy Crown Prince have signaled that they want change, but the needed change will not come about in the absence of political reform.

In Malaysia, the existence of massive inequality and ethnic discrimination, corruption and poor governance have led to socio-economic injustice and disintegration. Fragile institutions that were being established and nurtured could not withstand the onslaught of transgressions.

Some oil (gas)-exporting OIC countries not in our sample, such as Qatar and the UAE (especially the Emirate of Abu Dhabi), are so rich on a per-capita basis that they have more time and leeway to establish effective institutions. If today the rulers think that the current state of affairs can go on forever, they are sorely mistaken. The future of oil and gas is uncertain and the general citizenry may demand representative governance, the rule of law and more freedoms. Reforms and institution building are much more likely to succeed now before popular demands and unrest appear.

References

Abdul Khalid, M. (2014). *The colour of inequality: Ethnicity, class, income and wealth in malaysia.* Petaling Jaya: MPH Group Publishing Sdn Bhd.

Ahmad, Z. (2007). The Tragedy of May 13, 1969. Retrieved from http://www.malaysianbar.org.my/echoes_of_the_past/the_tragedy_of_may_13_1969.html

Caselli, F., & Cunningham, T. (2009). CEP Discussion Paper No. 913 March 2009 Leader Behavior and the Natural Resource Curse, (913).

Ebrahim, M. S., Girma, S., Shah, E., & Williams, J. (2014). Dynamic capital structure and political patronage: The case of Malaysia, *International Review of Financial Analysis*, 31, pp. 117–128.

Economist, The. (2016). Nothing to see here. *Asia*. Retrieved from http://www.economist.com/news/asia/21710280-billions-are-stolen-only-whistle-blower-goes-jail-nothing-see-here

Fraser, D. R. (2006). Capital structure and political patronage : The case of Malaysia, 30 (November 2016), 1291–1308. https://doi.org/10.1016/j.jbankfin.2005.05.008

Gale, B. (1981). Petronas Malaysia's National Oil Corporation. *Asian Survey*, *21*(11), 1129–1144. https://doi.org/10.2307/2643998

Jalal, T. S., & Bodger, P. (2009). National energy policies and the electricity sector in Malaysia. *ICEE 2009 3rd International Conference on Energy and Environment*, (December), 7–8. https://doi.org/10.1109/ICEENVIRON.2009.5398618

Khan, M. H. (2000). Rents, rent-seeking and economic development : An introduction. *Rents, Rent-Seeking and Economic Development: Theory and Practice in Asia*, 1–24. https://doi.org/10.1017/CBO9781139085052

Kit, G. M. (1998). Malaysia's political economy: Politics, patronage and profits. *Pacific Affairs*, *71*(4), 593–594. https://doi.org/10.1017/CBO9781107415324.004

Koon, H. P. (1996). Chinese responses to Malay hegemony in Peninsular Malaysia 1957–1996. *Southeast Asian Studies*, *34*(3), 500–523.

KRI. (2016). *Climbing the ladder: Socio economic mobility in Malaysia*. Retrieved from http://www.krinstitute.org/assets/contentMS/img/template/editor/SM_English.pdf

Mirakhor, A. (2009). Islamic economics and finance: An institutional perspective. *IIUM Journal of Economics and Management*, *17*(1:ProQuest Asian Business & Reference).

Ragayah, H. M. Z. (2008). Income inequality in Malaysia. *Asian Economic Policy Review*, *3*(1), 114–132. https://doi.org/10.1111/j.1748-3131.2008.00096.x

Shamsul, A. B. (2001). A history of an identity, an identity of a history: The idea and practice of "Malayness" in Malaysia reconsidered. *Journal of Southeast Asian Studies*, *32*(323), 355–366. https://doi.org/10.1017/S0022463401000194

· 7 ·

SUMMARY, CONCLUSIONS AND LOOKING AHEAD

Introduction

We first asked if our selected oil-producing OIC countries suffer from the resource curse and if so whether institutional weakness is the root cause? We presented a review of the resource curse theory, explanations and suggestions on how to reverse the curse. The landmark study by Sachs and Warner shows that none of the resource-rich countries with extremely abundant resources in 1970 grew rapidly for the next 20 years. The study of oil economies pointed to an even stronger negative association between growth and natural resource abundance. The study triggered further research in the area of the resource curse. The gap in the existing literature analysis indicates that there were few studies linking the OIC cluster countries to the resource curse. Moreover, there was no evidence that research had looked into Islamic principles and teaching as a potential solution to address the curse and manage resources. Our contribution to policy making is to develop recommendations that address the curse in totality and according to the teachings advocated by Islam.

Summary and Conclusions

A cross-country growth model, typical in the resource curse literature, was used as the core model. In line with the work of Latif, Farooq, Loganathan, and Shahbaz (2014), El Anshasy and Katsaiti (2013), Bjorvatn, Farzanegan, and Schneider (2012) and Alexeev and Conrad (2011), economic growth was the dependent variable and oil rent represented the preferred resource indicator. Static panel methods of fixed effects and random effects were used as part of the testing procedure. The PMG and MG methods were used to supplement the findings for robustness. The results indicate that resource curse is present in the sample study that includes ten selected OIC resource-rich countries.

The identified root cause and transmission mechanism are seen in the low quality of institutions. This led to extensive discussions that centered on institutions in Chapter 3. We examined the conventional viewpoint on the definition of institutions and its relation to economic growth and the resource curse. The conventional literature views institutions as a set of rules and norms that are meant to constrain the behavior of individuals in the interests of maximizing wealth. The actions of humans and the economic conditions they face shape the context and dimensions of institutions. The capitalistic self-interest notions driving institutional value were traced back to the period of the Protestant Reformation. However, there is a growing body of literature that regards institutions as critical to economic growth and the resource curse. Most of the research indicates the need to address institutional weakness in order to address the curse.

OIC membership is based on one or more of the following; (1) governments who have adopted Islam as the official state religion, (2) Islam as practiced as a primary religion, (3) the country being made up of significant Muslim population, or (4) self-declaration as an Islamic republic (Rehman & Askari, 2007). Structural commonality motivated the discussion in the chapter in which the definition of institutions according to Islam was presented. Institutions according to Islam differed in source and objectives. Institutions in Islam are based on rules of Allah (swt) and do not aim to restrain man. Rather they are aimed to drive men to a certain behavioral conduct that will guide them to their best potential.

The Islamic institutional economic framework takes the form of the Archetypal Model framework that the Prophet (sawa) operationalized during the Medinan period. It represented successful implementation of the

Metaframework rules of the Creator and is the best type of model to govern economic, social and political affairs of the community. The model and its elements that govern sanctity of contract, property rights, trust, governance, existence of markets, production and distribution of resources, income and wealth, behavior of market participants and post-market distribution were presented and discussed at length to reflect upon its completeness and how it served to achieve economic, political and social justice. The state of institutions in the countries of our study were reviewed and juxtaposed against the stated visions of Islam. The countries did not have a high Islamicity index, had poor corruption ratings and did not manage their depletable natural resources in line with providing equal benefits to all citizens of all generations.

The recommendations proposed in Chapter 5 first suggested for the redirection of education towards understanding the *wālayah* of Allah (swt) and to the rules of the Archetypal Model. This redirection aims to ensure that all Muslims understand their relationship with the Supreme Creator and internalize the ordained rules so that they may orient their daily lives in full rule compliance. Rule compliance provides self and collective benefits and curbs counterproductive behaviors that typically cause institutional weaknesses and, in turn, the ineffective management of natural resources. Through being cognizant and engaging to internalize the rules, men will arrange their transactions, interpersonal dealings and economic affairs according to the rules of Allah (swt). Past research indicates that natural resources are typically governed under state ownership and the way the state is governed seldom escapes national and diplomatic interests (Robinson, Torvik, and Verdier, 2014). The duties and accountabilities of the state and its citizens were elucidated according to the Islamic paradigm where the duty to ensure social justice lies with all members of the society. State authorities do not have the free will to conduct the matters of the state according to their preference. Rather they hold the profound responsibility to ensure that the rules of Allah (swt) are preserved. Thus, state representatives are selected by the demonstration of their esteemed position of rule compliance that is higher than that of the general population and their commitment to uphold the rules. The order of balance in resource allocation also lies in the state as they oversee the management of resources according to the rules of Allah (swt) at the macro level.

The responsibilities held by citizens are of equal import. They are to remain vigilant behind the ruling authority for as long as the office can demonstrate a high degree of compliance to the rules of Allah (swt). Citizens are also expected to keep an eye on the conduct of the authority and stand up for members of society in any instances of injustice. In extreme violations, citizens are given avenues to express their support. The rules of Allah (swt) governing both state and citizens are all encompassing and are meant to curb corrupt practices, waste, opulent spending, rent seeking, patronage and abusive power plays relating to the institutional issues commonly associated with lax in management of resources. The final recommendation refers to enhanced supervision, based on practices from the historical Medinan period of the Prophet (sawa). This enhanced supervision takes the form of *muhtasib* at the national level and among countries. It is aimed at fostering social solidarity irrespective of racial and national relations.

When we took a more in-depth look at two countries in our study, namely, Saudi Arabia and Malaysia, we found that the corrosion in quality of institutions has resulted in misallocation of resources and imbalanced socio-economic conditions. In the case of Saudi Arabia, the central impediment to progress and establishing effective institutions is the mindset of the autocratic rulers. The Al-Sauds believe that the country is theirs as their ancestor took it over with the sword. Thus, they have been reluctant to secede power and control of the national treasury to the people. In the case of Malaysia, a Muslim country that most observers would have earlier selected as one of less than a handful of successful resource-rich OIC countries, is today under serious strain. Ethnic discrimination and corruption have increased. In the case of Saudi Arabia, institution building has hardly begun; yes, there are schools and universities but the bedrock institutions of law and representative governance are not even on the horizon. In Malaysia, institution building had started but it was still in its early development stages and fragile; and today it is being overwhelmed by ethnic discrimination and corruption. We took a very brief look at Qatar, a country that is not in our sample of countries, because it presents different conditions than what we see in Saudi Arabia and Malaysia. Qatar's vast riches may afford the rulers more time and options for reforms, but even they should begin now before it becomes too late.

If institutional weakness continues to go unaddressed in oil-producing OIC countries, they will continue to tread down a path of corruption, waste, underdevelopment and ultimately internal conflict that could even spill into civil war. Operating within weak institutional scaffolding results in unbalanced order and unjust societies, which invariably becomes more glaring when enormous oil revenues flow into government coffers but are either absconded or wasted. The impact of injustice and underdevelopment invariably has the severest impact on minorities, the poor and more generally on the weakest groups of society. CIC countries have a collective responsibility to stop this transgression from progressing further by taking immediate remedial actions. The Islamic institutional framework offers the most complete, immutable and just solution for countries that profess Islam. The members of these Muslim communities need to look no further than the divine teachings of the Qur'an and the Sunnah of the Prophet (sawa) for sustainable solutions to ameliorate the resource curse problem while they still enjoy abundant revenues from the sale of oil and gas and have many options looking ahead.

References

Alexeev, M., & Conrad, R. (2011). The natural resource curse and economic transition. *Economic Systems*, *35*(4), 445–461. https://doi.org/10.1016/j.ecosys.2010.10.003

Bjorvatn, K., Farzanegan, M. R., & Schneider, F. (2012). Resource curse and power balance: Evidence from oil-rich countries. *World Development*, *40*(7), 1308–1316. https://doi.org/10.1016/j.worlddev.2012.03.003

El Anshasy, A. a., & Katsaiti, M.-S. (2013). Natural resources and fiscal performance: Does good governance matter? *Journal of Macroeconomics*, *37*, 285–298. https://doi.org/10.1016/j.jmacro.2013.05.006

Latif, S., Farooq, A., Loganathan, N., & Shahbaz, M. (2014). Empirical evidence on the resource curse hypothesis in oil abundant economy. *Economic Modelling*, *42*, 421–429. https://doi.org/10.1016/j.econmod.2014.07.020

Rehman, S. S., & Askari, H. (2007). An Economic Islamicity Index. *Global Economy Journal*, *7*(2), 3–4.

Robinson, J. a., Torvik, R., & Verdier, T. (2014). Political foundations of the resource curse: A simplification and a comment. *Journal of Development Economics*, *106*, 194–198. https://doi.org/10.1016/j.jdeveco.2013.09.004

INDEX

A

B

C

D

E

F

G

H

I

N

O

P

Q

R

U

V

W

Y

Z

ECONOMIC, POLITICAL AND SOCIAL INSTITUTIONS IN ISLAM

Hossein Askari and Dariush Zahedi
Series Editors

This series gathers research focusing on the idea that a prerequisite to successful actions to improve economic performance is "a viable polity that will put in place the necessary economic institutions and provide effective enforcement." Some—acknowledging that institutions are critical to any explanation of economic development—question why similar institutional structures produce different results in different countries. Others have concluded that what matters for good economic performance is how well institutions match their settings and how flexibly they adapt to changes as opposed to a specific design. This series aims to curate research from scholars in the fields of economics, politics, technology and the humanities to gather interdisciplinary perspectives on institutional policy throughout the Islamicate world.

For additional information about this series or for the submission of manuscripts, please contact:

Peter Lang Publishing
Acquisitions Department
29 Broadway, 18th floor
New York, NY 10006

To order other books in this series, please contact our Customer Service Department:

800-770-LANG (within the U.S.)
212-647-7706 (outside the U.S.)
212-647-7707 FAX

Or browse online by series at:

www.peterlang.com